Guido Küstel

Nevada and California Processes of Silver and Gold Extraction

Guido Küstel

Nevada and California Processes of Silver and Gold Extraction

ISBN/EAN: 9783337338817

Printed in Europe, USA, Canada, Australia, Japan

Cover: Foto ©berggeist007 / pixelio.de

More available books at **www.hansebooks.com**

NEVADA AND CALIFORNIA

PROCESSES

OF

SILVER AND GOLD EXTRACTION,

FOR GENERAL USE, AND ESPECIALLY FOR THE

MINING PUBLIC OF CALIFORNIA AND NEVADA,

WITH FULL EXPLANATIONS AND DIRECTIONS FOR ALL METALLURGICAL
OPERATIONS CONNECTED WITH SILVER AND GOLD FROM A
PRELIMINARY EXAMINATION OF THE ORE TO
THE FINAL CASTING OF THE INGOT.

ALSO,

A DESCRIPTION OF THE GENERAL METALLURGY OF SILVER ORES.

BY

GUIDO KÜSTEL,

Mining Engineer and Metallurgist, former Manager of the Ophir Works, etc.

Illustrated by accurate Engravings.

SAN FRANCISCO:
FRANK D. CARLTON.
1863.

Entered according to Act of Congress, in the year of our Lord 1863, by

FRANK D. CARLTON,

In the Clerk's office of the District Court of the United States in and for the Northern District of California.

TOWNE & BACON, PRINTERS, EXCELSIOR OFFICE,
No. 536 Clay Street, opposite Leidesdorff.

PREFACE.

The rapid extension of Silver Mining enterprise, in consequence of numerous discoveries of rich and extensive silver-bearing lodes in California, and especially in Nevada Territory, has excited a general desire for information of such methods of extracting Silver and Gold from the different classes of ores, as are practical and adapted to our circumstances.

In representing the different modes of working silver ores, it appeared a necessity in some cases to treat the theory fully, as for instance concerning the roasting process, which it is impossible to carry on for certain purposes, if the mutual action of the various substances, under the influence of heat, is not well understood. In regard to this point I am indebted for valuable information to "F. Plattner, on Roasting Processes."

I have prepared the text of this work with care, and supplied the first part with distinct and detailed drawings, so that one by studying it can learn to perform all the operations. If, however, there are yet many things not satisfactorily explained, or not thoroughly treated, it may be attributed to the circumstance that our metallurgy is very young.

The second part contains a description of the general metallurgy of silver ores, for which I am indebted to Kert's Metallurgy, one of the latest and best German works. It is not likely that melting or precipitation will ever be introduced in California or

Nevada for general use, yet, under particular circumstances, in one case or another, it may in future prove useful to adopt processes, which are not practiced here at present. For this and other reasons a brief but careful description of the general metallurgy of silver ores is given.

The assaying of ores, tailings, bullion-metal, etc., is so closely connected with the mill and mining business that tables showing the amount of fine metal per ton of ore and also the value of silver and gold per ounce at different degrees of fineness have been added for reference.

DAYTON, N. T., May, 1863.

TABLE OF CONTENTS.

PART FIRST.

Chapter I.

	Page.
Remarks on the Blowpipe	9
Blowpipe Tools	10
Blowpipe Reagents	10
Blowpipe Instruments	12
Blowpipe Materials	14
Blowpipe Use	15
Blowpipe Examination on Charcoal	16
Blowpipe Examination on Charcoal with Soda and Borax	18
Blowpipe Examination in a closed Glass Tube	20
Blowpipe Examination in an open Glass Tube	22
Systematic Procedure for the Determination of Gold and Silver Minerals	23
Analysis of retorted Amalgam or Bullion	26
Blowpipe Assay	28
Specific Gravity	33
Hardness	35

Chapter II.

DESCRIPTION OF GOLD ORES.

	Page.
Gold	36
Gold in Combination with Silver	36
Gold in Combination with Tellurium	37
Gold in Combination with Tellurium and Lead	37
Gold in Combination with Mercury and Silver	37

DESCRIPTION OF SILVER ORES.

Native Silver	38
Silver Glance (Sulphuret of Silver)	38
Stromeyerite (Silver-copper Glance)	39
Sternbergite	39
Sternbergite of Gold Hill	39
Brittle Silver Ore	40
Polybasite	40
Miargyrite	41
Dark Red Silver Ore (Ruby Silver)	41
Light Red Silver Ore (Ruby Silver)	42
Xanthocone	42
Silver Fahlerz (Argentiferous Gray Copper Ore)	42
Horn Silver (Chloride of Silver)	43
Embolite (Chlorobromide of Silver)	44
Bromyrite	44
Iodyrite	44
Iodid of Mercury and Silver	45
Antimonial Silver	45
Naumannite (Selenid of Silver)	46
Eucairite	46
Hessite (Tellurid of Silver)	46
Bismuth Silver	47
Silver Amalgam	47
Arquerite	47

Chapter III.

Fire Assay	48
Tools	48
Materials	49
Gold and Silver Assay	50
Lead Assay	57

Chapter IV.

Extraction of Gold	59
Remarks	59
Amalgamation in the Battery	59
Amalgamation on Copper Plates	61

CONTENTS.

	Page
Amalgamation in **Arrastras**	61
Amalgamation in Pans	63
Extraction of Gold **by Chlorination**	64

CHAPTER V.

Extraction of Silver	67
Division of Processes	67
Wet Process	67
Chemicals	69
Sulphate of Copper	69
Sulphate of Iron	71
Bisulphate of Soda	71
Alum	71
Sulphuric Acid	71
Salt, Common	72
Chloride **of Copper**	72
Subchloride of Copper	73
Protochloride of Iron	73
Chloride of Iron	74
Chemicals, quantity per Ton **of Ore**	74
Amalgamation in **Pans**	76
Amalgamation in **Wheeler's Pans**	81
Roasting Process	90
Behavior of Chlorine	94
Behavior of Hydrochloric **Acid**	96
Behavior of Salt	96
Roasting, remarks	98
Roasting for the Barrel and Veatch's Amalgamation	101
Roasting for Pan Amalgamation	106
Roasting of Silver Ores rich in Antimony	110
Roasting in a Mechanical Furnace	114

	Page
Loss of Silver in Roasting	115
Amalgamation in Barrels	117
Amalgamation **in Veatch's Steam** tubs	122
Amalgamation **in** Pans	124
Cold Process	128
Patio or American Heap Amalgamation	128
Retorting	132
Melting	134
Assay of the Bar	143
Melting Process	145
Melting	154
Mixture of Ore and Fluxes	151
Separation of Lead and Silver	154
Refining	162
Pattinson's Process	164

CHAPTER VI.

Description of the Common Pan	169
Description of Wheeler's Pan	170
Description of Wheeler's Agitator	173
Description **of** Hepburn & Peterson's Pan	174
Description of Roasting Furnace	176
Description of Mechanical Roasting Furnace	177
Description **of** Retort	178
Description **of** Crucible Furnace	179
Description **of** the Melting Furnace	180
Description **of** the Cupelling Furnace	183
Description of the Refining Furnace	186

PART SECOND.

GENERAL METALLURGY OF SILVER ORES.

CHAPTER I.

Silver Ores, extensively used **for Extraction**	189
Argentiferous Lead **Ores**	190
Argentiferous Copper Ores	190
Argentiferous **Zinc Ores**	191
Argentiferous Iron and Magnetic Pyrites	191
Assays	191
Dry or Fire Assays	192
Assays of Rich Ores	193
Assays of Poor **Ores**	196
Assays of Argentiferous **Alloys**	197

Wet Assays	199
Blowpipe **Assays**	199

CHAPTER II.

Methods of extracting Silver	205
Melting with Lead or Lead Ores	205
Amalgamation	207
Dissolution and Precipitation	208
Division of principal Methods of Extraction	211

CHAPTER III.

Extraction of Silver in the Dry Way. 216

CONTENTS.

	Page.
Extraction with Lead	216
Production of Argentiferous Lead	220
Melting in Crucibles	220
Melting in the Cupelling Furnace	221
Melting with Unroasted Lead Ores	222
Melting of Argentiferous Lead Ores	223
Melting of Argentiferous Copper Ores	224
Lead Manipulation with Matts	225
Concentration of Silver in Matt	225
Extraction from Matts by Lead	228
Liquation	231
Capellation	235
Cupellation in unmovable hearths	235
Cupellation in movable hearths	243
Pattinson's Process	244
Parke's Process	249
Refining of Silver	251
Refining on Movable Tests	254
Refining under Muffles	255
Refining at the Mint of Clausthal	257
Refining in Reverberatory Furnaces	257
Refining in **Crucibles**	259
Extraction of **Silver in the Wet Way.**	261
Amalgamation in Barrels	262
Roasting	263
Amalgamation of **Ores**	265
Amalgamation of Copper **Matt**	267
Amalgamation of Speiss	268
Amalgamation of Black Copper	268
Augustin's Process	270
Augustin's Process on Matts	270
Ziervogel's **Process**	275
Patera's Process	279
Tables showing the Amount of Fine Metal per Ton of Ore	283
Tables showing the Value of Silver per Ounce in the Bar	297
Tables showing the Value of Gold per Ounce in the Bar	311

ERRATA.

Page 70, bottom line, *for* **30-100** *read* **30-1000**.
    72, line 1, *for* **condition. In** *read* **condition, in**.
    149, " 16, *for* **sulphurets** *read* **sulphates**.
    161, " 6, *for* **as off** *read* **off as**.
    163, " 20, *for* **can be** *read* **cannot be**.
    171, " 18, *for* **ledge** *read* **edge**.
    190, " 9, *for* **rich** *read* **richer**.
    192, " 3, *for* **fast** *read* **first**.
    203, " 14, *for* **block** *read* **black**.
    217, " 4, *for* **lead crucibles** *read* **lead in crucibles**.
    221, " 24, *for* **hesth** *read* **hearth**.
    231, " 20, *for* **silver containing** *read* **silver-containing**.
    241, " 16, *for* **silver** *read* **sides**.
    248, bottom line, *for* **crystallization** *read* **cupellation**.
    269, line 17, *for* **oxyd** *read* **mixed**.

CHAPTER I.

The knowledge of the composition of different classes of gold and silver ores is of great importance to every miner, especially to the millman. This knowledge is easily acquired by the aid of the blowpipe. I have endeavored to show the use of the blowpipe, limited to gold and silver ores, with as few and simple instruments as possible. The blowpipe is a very insignificant looking instrument; but a short acquaintance with it will show that no person engaged in mining can get along without it, no matter whether he be a good judge of ore or not. There are many different kinds of silver ores, the richness of which cannot be determined by the eye of the best judge.

Long study and extensive apparatus are necessary to learn the mode of making the wet analysis, and the application demands much time and patience in every instance. This proves unsuited to the general use of the millman. But the use of the blowpipe is acquired in a short time, and half an hour's blowing upon a piece of ore will tell us all about what it is, or how much silver it contains.

The instructions in regard to the use of the blowpipe in this book refer only to such gold and silver ores as are described herein.

APPARATUS AND RE-AGENTS

For Analysis of Gold and Silver Ores by the Blowpipe.

TOOLS.

Section 1. *A Blowpipe.*—A proper blowpipe must have a chamber to receive the condensed water, and should have a mouth-piece of horn or ivory, without which long blowing is very tiresome; a common alcohol lamp; a small agate or porcelain mortar with pestle; a magnifying glass; a small magnet; a small hammer and square piece of steel to serve as an anvil; a steel forceps; a small file and glass tubes. It is best to procure glass tubes in pieces two or three feet long, about three-sixteenths of an inch in the clear. They are cut with the file into pieces four and eight inches long, of which the shorter ones are ready for use, being open at both ends. The eight-inch pieces are held in the alcohol flame with the fingers at each end. When the tube gets red-hot in the middle, it will part into two pieces by drawing, forming tubes closed at one end.

RE-AGENTS.

Sec. 2. 1. *Carbonate of Soda.*—It makes no difference whether carbonate or bicarbonate of soda is used, but it must be free from sulphuric acid. To ascertain the absence of sulphuric acid, a small portion of the soda may be melted with the blowpipe on charcoal, till it draws into the coal. This part is then broken out with a penknife, put on a blank spot of a silver coin, and moistened with water. After one or two minutes it is removed. If the soda contains any sulphuric acid, a yellow or brown spot will be seen on the silver. In this case the soda must be purified, which may be done in the following way: The pulverized bicarbonate is introduced into a glass funnel on a piece of filtering paper, double folded, under which some loose cotton was placed. Cold distilled water is poured over it several times, till the soda is sufficiently washed. This can be ascertained by trying a little on charcoal, as before described. This soda is used for the detection of small quantities of sulphur.

2. *Borax.*—The borax of commerce is generally pure enough to serve the purpose. A small piece, melted on charcoal, if pure, will turn into a colorless glass pearl. If it appears colored, the borax must be purified, by dissolving it in hot water, and subsequently crystallizing it. The borax has the property to dissolve oxyds of metals when melted with them, assuming different colors. By this treatment, several metals are recognized.

3. *Bisulphate of Potassa.*—This is easily prepared. One ounce of sulphate of potassa, roughly ground, and half

an ounce of sulphuric acid, are heated in a porcelain cup by the alcohol lamp till all is in a quiet, transparent fluid condition. The cup is then taken from the fire, and the contents poured on a piece of sheet iron. The bisulphate of potassa is used in testing ore for iodine and bromine.

4. *Tin.*—The tinfoil is used to discover a small quantity of copper in the ore. The tin has the property of reducing oxyds of metals to a lower state of oxyd in glass fluxes, effecting thus a different appearance, by which the metal can be recognized more easily.

5. *Oxyd of Copper.*—A piece of clean copper is dissolved in nitric acid, evaporated to dryness, and heated by degrees in a small porcelain cup to red heat. It is used for the detection of chlorine.

APPARATUS AND RE-AGENTS

For Blowpipe Assays.

INSTRUMENTS.

Sec. 3. 1. *An Assay Balance.*[*]—This is the most expensive necessary instrument. The best for our purpose is the blowpipe balance, which answers also perfectly for the fire and bar assays. Such a balance costs from one hundred to one hundred and twenty dollars.

---

[*] Good ones are made by W. Schmoltz, in San Francisco, from seventy-five to eighty dollars.

The weights should be grain weights (Troy). Ten grains represent the unit, which is divided into: 1,000, 400, 300, 200, 100, 40, 30, 20, 10, 4, 3, 2, 1 thousandths.

A little spoon, of the size of an ear-pick, made of ivory or copper, to weigh out the powdered ore. The handle of this spoon should have the form of a spatula, for mixing ore with lead and borax glass in the capsule.

A small wooden solid cylinder, to make paper tubes, in which the assay is packed and melted (Fig. 1 *a*). The paper tubes are made best of fine letter paper, which is cut into strips one and one-half inches by one. The strip (Fig. 1 *b*) is rolled round the cylinder, so that the long side of the paper may project beyond the end of the cylinder, the distance of its diameter. The projected part is pressed down where the strip ends with the aid of the handle of the small spoon, then the left and right side, and finally the remainder. The cylinder is then turned, and the folded end knocked on the table two or three times. The tube is taken off, and is ready to be used.

A drill to bore holes in the charcoal, in which the paper tube with the ore is placed to be melted. This drill (Fig 2) has the shape of a cross, or a double chisel crossing at right angles, and must be a little larger than the cylinder (Fig. 1).

A brass capsule, of the shape as represented in Fig. 3. It is two and one-half inches long. At the line *a* one-eighth of an inch deep, and one-fourth of an inch wide. At the line *b*, one-fourth of an inch deep, and seven-eighths wide. This capsule serves for mixing ore with fluxes.

A cupel holder (Fig. 4 *a*), is made of copper or brass; its size according to the size of the cupels.

Cupels (Fig. 4 *b*), made of boneash. The smallest kind will answer.

MATERIALS.

SEC. 4. 1. *Borax.*—The borax for blowpipe assays is used as borax glass. A black lead or other crucible is made red-hot, and some borax introduced. It will swell up at first, and then melt down. The borax is added in small portions, till the required quantity is obtained. The fluid borax is poured on a clean iron plate, pulverized, and kept in a bottle with a ground glass stopple. If exposed to the air, it absorbs moisture, and is unfit for our purpose.

2. *Lead.*—The lead must be perfectly free from silver. Pure lead can be obtained in the following way: Acetate of lead is dissolved in hot water and filtered. A piece of clean zinc will precipitate the lead. This must be washed several times with clean water, in order to get rid of the zinc, which is now in solution. The lead is then dried and melted in a clean crucible with some soda and borax. Small quantities of lead cannot well be granulated; it is then better to file it.*

---

* Scales, balances, tools, assay apparatus, and all chemicals mentioned in this work, are to be found at John Taylor's, Washington Street, San Francisco.

USE OF THE BLOWPIPE.

Sec. 5. The only difficulty which some beginners find in using the blowpipe is the production of a continual blast. To effect this, one must blow the air by aid of the cheeks, and supply the mouth with air by the lungs. The breathing must continue through the nose during the blowing. The direct use of the lungs for the blowpipe must be avoided, as it injures the chest.

The flame shows two different parts of different character. If the blowpipe is kept so that the point touches the flame of the candle, as represented in Fig. 5 *a*, it produces a yellow inner flame, which has the property of preventing oxydation, on account of its amount of carbon. It effects the reduction of many oxyds of metals, and is called the reduction flame. It must be used so as to cover the test particle (Fig. 5 *b*).

The other, the outer flame, is produced by holding the blowpipe point a little into the flame, above the wick (Fig. 6 *c*). This outer flame is blue, sharp-pointed, and must be free of yellow parts. It has the reverse property of the inner flame, oxydizing the metals when before the blue point (Fig. 6 *e*).

The wick of the candle must be kept short, and the end bent down like a hook. Particles of thread or other matters on the wick are removed, because they interfere with the purity of the flame.

EXAMINATION ON CHARCOAL.

SEC. 6. In examining the ore on charcoal, attention must be paid first to the odor, then to the coating by the volatile metals or oxyds of metals. The grain, taken into examination must be small, about the size of the button represented by *e* in Fig. 6. The charcoal must be sawn lengthwise. On a flat surface a small hole is made near the edge with a penknife, to receive the particle of ore intended for the test.

The ore may be blown at, first with the outer or oxydation flame. If the ore decrepitates, it must be pulverized in the agate mortar. As soon as the test gets red hot, the odor must be observed. Sulphur emits sulphurous acid, a well-known odor. Selenium has a disagreeable smell, somewhat like rotten radish, and the arsenic is recognized by the garlic-like odor. If sulphur is present, a small amount of arsenic could not be detected. In this case the reduction flame is applied, after no more odor of sulphurous acid is observed, and recognized by the odor.

Next, the attention should be directed to the coating of charcoal by volatile combinations. It is of importance to notice the distance of the coating from the test, and also the color, which generally changes on cooling. Some coating when played upon with the oxydation flame, will move from one place to another, or will cause a colored halo, when treated with the reduction flame. It is necessary, therefore, to get acquainted with the different coatings produced **by** different metals, **not only those** composing the **gold and silver ores,** described in this book, but also others, which appear sometimes **in the** silver ores, though **not** essential **to their** classification.

Lead—Gives a coating **with both flames near the** assay. It appears dark lemon **yellow when hot, changes into** sulphur **yellow when cold. In thin layers it appears** bluish **white. The coating may be driven away by** both flames, **coloring the reduction flame blue. If iodine be present, the** coating is **far off from the test, intensely green, or** light green if lead **predominates, with a yellow seam inside. On** cooling it changes **very little.**

Antimony.—It **covers the charcoal** with a white oxyd or bluish **white if in a thin layer.** The coating **is not** close to the assay **and is** easily driven away. **If a great deal of silver be present, the coating appears pink.**

Zinc.—The coating of **the oxyd of zinc is** near **to** the test, appears yellow when **hot, and** turns **white** on cooling. Heated **with the oxydation flame,** the coating becomes luminous.

Tin.—The coating is quite **close to the test. It is yellow when hot** and white when cold, resembling **the coating of zinc. Becomes also luminous. In the reduc-**

tion flame the coating of oxyd can be **reduced** to the metallic state.

Selenium—Gives **a steel gray** coating of a feeble metallic lustre, showing somewhat **violet** color further off; **played** upon with the flame, **it emits a** rotten radish-like odor.

Tellurium.—It **fumes and coats the coal white** with a red or dark **yellow seam.**

Examining **ore on charcoal, it cannot be expected to get always coatings which agree exactly** with the description, **there being generally** different metals present. **If, for instance, the ore** under examination contains lead, **zinc, and antimony, the** white coating with **a blueish seam will appear** first, and is recognized as antimony. **But it soon** becomes yellow from the oxyd **of** lead, which, **mixed with the** white antimony coating appears lighter. The zinc **coating which is** also yellow colored **is not** distinguishable **at first** from the lead coating, but this **latter can be blown** away with the oxydation flame, leaving **the zinc oxyd,** which **appears** luminous when **played upon with the** oxydation **flame.**

EXAMINATION WITH BORAX AND SODA,
On Charcoal.

Sec. 7. The borax **assumes a** globular shape when melted on charcoal with **the** blowpipe. If pure, it appears colorless, and dissolves the oxyds of metals. If sulphurets are taken in examination, they must be powdered and a small portion of it treated with the reduction flame, on charcoal, in order to drive out the sulphur and arsenic. If the sulphurets melt, the powder may be mixed with charcoal dust and treated to a dark red heat. A very small part of this roasted ore, or not roasted if the test consists of oxyds, is put on the borax pearl, and melted with the oxydation flame. If the color of the glass is not perceptibly changed, another particle of the ore can be dissolved, but it is never advisable to take too much in the first instance. If the globule be too dark in color, it may be flattened by a pair of pincers after blowing, before getting stiff, and then observed against the light.

A colored borax glass appears often different when cold or warm, also if treated with the outer or inner flame; there are, however, but a few metals to be considered for our purpose.

Iron.—It colors the borax **reddish yellow or reddish brown,** when hot and played upon with the oxydation flame. It turns **yellow, or colorless if in a** small quantity, **when cold. A good reduction flame makes the** glass **reddish brown, but it becomes bottle green when cold.**

Copper—When hot, colors the glass green; when cold, blue. The presence of other metals renders these signs uncertain. The surest way is, to reduce the oxyd of copper which is dissolved in the borax, to a suboxyd by which the glass assumes a light red color, becoming opaque.

When the well-calcined test is dissolved in borax, a small piece of tinfoil is laid on the pearl and blown with the reduction flame. If there is only a small quantity of copper in the ore, the borax pearl will appear brick red and opaque when cold. If the color is grayish brown, there is also some antimony in the test. If the pearl becomes black there is a great deal of antimony, which must be driven off by a better roasting.

EXAMINATION IN A CLOSED GLASS TUBE.

SEC. 8. A particle of ore, of the size as represented in Fig. 7 a, is introduced into a glass tube (Fig. 7 b) and heated by an alcohol lamp. The heat must be raised gradually to redness, also increased, if no sublimate appears, directing the alcohol flame by the blowpipe on that part of the tube where the particle of ore is, till the glass commences to melt.

The volatile substances sublimate on the cooler part of the tube (Fig. 7 c), assuming different colors, thus enabling us to recognize the substance.

a. A dark yellow or reddish-brown sublimate when hot, turning the sulphur yellow when cold, indicates sulphur. Only such sulphurets which contain two atoms of sulphur to one atom of metal emit sulphur in a closed tube.

b. A reddish yellow **sublimate** which appears at a high temperature under the blowpipe, becoming sulphur yellow when cold, shows iodine.

c. A dark reddish-brown, **almost black** sublimate, becoming red or reddish-yellow after cooling, shows **the** presence of arsenic and sulphur.

d. A **black sublimate,** quite close **to the test, arising by the aid of the** blowpipe and turning **brownish-red when cold, indicates** sulphur and antimony.

e. A **black** sublimate **without lustre,** rendering a red powder **when scratched, indicates sulphur and** mercury.

f. A grayish-white sublimate, consisting of transparent **crystals, which** can be distinguished **under the** magnifying glass, indicates antimony. Some ore will not yield this sublimate, **unless it is treated with the blowpipe.**

g. A reddish **black sublimate,** obtained **by** the aid of the blowpipe, giving a dark red powder, indicates selenium, **which can be observed** at the open **end,** by the odor of rotten radish.

h. A **gray sublimate** indicates mercury. **The quicksilver globules can be** perceived **by** a magnifying glass.

EXAMINATION IN A GLASS TUBE,
Open at Both Ends.

SEC. 9. For the purpose of examining the ore in an open tube, the test is introduced so as to have it near one end. That part of the tube is heated over the alcohol flame, while the other end is raised a little, so as to create a draft. It is often required to raise the heat by aid of the blowpipe, as described in Section 8. The ore is used in shape of grain, about as large as hempseed; but, if it was observed to decrepitate in the closed tube, it must be pulverized, and in that state introduced into the tube.

There are substances or combinations, which are not volatile in a closed tube, but being in contact with the oxygen of a draft of air in an open tube, they oxydize and escape, either in form of gas, when they can be perceived by their odor, or by their action on a strip of moistened blue litmus paper, which is placed in the upper part of the tube, or they sublime in the cooler part, nearer to the test, or further off, according to the degree of their volatility.

Care must be taken to raise the heat gradually, otherwise it may happen that a great part of the substance will sublime without change. There are several substances, which, treated in an open tube, can be recognized with certainty. Such substances are principally:

Sulphur.—Sulphurets, being heated in a tube, emit

sulphurous acid. The sulphur can be detected by the odor at the upper end of the tube, also by the litmus paper, which changes its blue color into red.

Antimony.—Antimony yields white fumes, condensing in the upper part of the tube to a white sublimate. The presence of lead makes the sublimate light yellow.

Selenium.—Selenium of silver emits at the upper end of the tube the odor of rotten radish.

Iodine.—Iodide of silver, if strong heat is applied, gives a yellow sublimate. It changes sulphur yellow when cold.

Quicksilver.—Quicksilver gives a sublimate of metallic mercury of a grayish-white appearance. If sulphur is present and strong heat is applied, a black coating will be formed.

SYSTEMATIC PROCEEDING

For Determination of Gold and Silver Ores.

Sec. 10. The use of the following systematic proceeding can be understood easily by an example:

A silver mineral, for instance, approved as such by examination on silver, must be observed first as to what lustre it shows, or whether it is dull. Suppose, then the mineral has a metallic lustre (see I). The color must be observed **next and** compared with those under **I**. The **ore is further** found to be "lead gray." We have **then to** proceed from the indicated letter **B** on **the right side to B on** the left, **and** examine accordingly, **whether** the mineral **gives a sublimate or** not. If, for instance, no sublimate **has been obtained,** we must proceed to *c* as indicated. **On the described** examination under *c* **the mineral appears tough, it can be cut** with a **knife. We go over to** Section **16, and see the** numbers **2 and 20, Silverglance and Hessite. The** description **of both will lead to the right determination** of the **mineral.**

I. LUSTRE–METALLIC OR SUB–METALLIC.

Color—white, grayish white, yellowish **white, or** yellow, see... *A*
Color—**lead-gray, blackish** lead-gray, or iron-black.......... *B*
Color—light **steel-gray**............................... *C*
Color—reddish **lead-gray**............................. *D*
Color—pinchbeck **brown**.............................. *E*

II. LUSTRE–RESINOUS AND ADAMANTINE.

Color—pearl-gray, yellowish green, **green, olive-green, lemon yellow or light yellow,** see..................... *F*

III. DULL.

Color—red, **dark red (sometimes** externally lead-gray), see... *G*
Color—blackish **blue**................................ *H*
Color—greenish black .... *I*

A It can be cut with a knife, see.......................... *a*
A Can not be cut, is **brittle**............................. *b*
B In a closed tube, **no** sublimate, even under the blowpipe.... *c*
B It gives a sublimate, with or without the blowpipe.......... *d*
C In a **closed tube,** no sublimate......................... *e*
C It **gives a** sublimate................................... *f*
D **In a closed** tube, dark red, **sublimate.** See Sec. 16 (8 or 9).
E **In a** closed or open tube, **no** sublimate. Sec. 16 (4).
F In a closed or open tube, no sublimate................... *g*
F Gives, with aid of the blowpipe, **a slight** sublimate........ *h*
G In a closed tube, **red-brown or reddish yellow sublimate.**
 Sec. 16 (10).
G It **gives three sublimates, black,** yellow, and **gray.** Sec. 16
 (16).
H In an open or closed tube, no sublimate. Sec. 16 (3 *a*').
I In an open or closed tube, no sublimate. Sec. 16 (11 *a*').
a It melts **on charcoal to a metallic white** globule. Sec. 16 (1).
a It **melts on charcoal to a metallic yellow or** yellowish globule.
 Sec. 15 (1).
b It melts **on charcoal** to a globule of metallic lustre, **coating the**
 coal white. **Sec.** 16 (17).
b It decrepitates somewhat, giving, **before fusing, a slight, very**
 volatile whitish **coating.** Sec. 15 (4) or Sec. 16 (22 or 23).
c It **can be cut** with a knife. Sec. 16 (2 or 20).
c It **can not be cut, is brittle.** Sec. 16 (3 or 4, *a* or 6), or Sec.
 15 (2).
d In a closed **tube it gives a reddish yellow** sublimate. Sec. 16
 (6 *a*').
e **On charcoal it fuses,** giving a yellow and white coating. Sec.
 15 (3).
f In a **closed tube, by** aid of the blowpipe, **a** dark red sublimate.
 Sec. 16 (11).

g On charcoal it fuses, emits an acrid odor, and leaves globules of silver; in a closed tube with bisulphate of potassa, emits no colored vapors. Sec. 16 (12).

g It gives, with bisulphate of potassa, red-brown vapors. Sec. 16 (13 or 14).

h In a closed tube, with bisulphate of potassa, violet vapors. Sec. 16 (15).

ANALYSIS OF RETORTED AMALGAM,

Or Bullion Metal.

Sec. 11. Black amalgam is first tried with the magnet on a small particle. If attracted, there is a great deal of iron in the amalgam. If it does not follow the magnet, it may still contain iron, which must be examined in a different way.

A small piece is introduced into a closed tube and heated by the alcohol lamp to redness. It generally gives out a whitish sublimate, consisting of minute globules of mercury, which can be detected by the magnifying glass, or by a piece of flattened gold, which, when introduced into the tube and rolled over in the sublimate, gets a coating of quicksilver.

Heated in an open tube, it sometimes colors litmus paper red. This proves the presence of sulphurets in the amalgam. Under such circumstances a black sublimate appears, consisting of sulphur and mercury.

A small particle of bullion metal or retorted amalgam is laid on charcoal and heated with the oxydation flame.

A slight yellow coating indicates lead; a bluish-white, antimony; or the coating is yellowish, and further off bluish white, proving the presence of both lead and antimony.

On a clean spot of the charcoal, some borax is melted into a transparent globule. The metal is placed against the borax and played upon with the reduction flame for half a minute. The oxydation flame would oxydize the silver, and color the pearl white enamel-like in cooling. The metal globule, when yet hot, is seized by a pair of pincers and taken out, leaving the borax pearl on the coal. To this pearl is added some tinfoil of the size of a small pinhead, and blown again only a short time with the reduction flame. If the pearl turns gray enamel-like or black, it shows the presence of antimony, which, on account of its small amount, could not have been perceived as a coating on the charcoal. In this case another small particle of the same test must be played upon with the pure oxydation flame for about one minute, in order to get rid of the antimony, and then treated with borax and tinfoil as above described. If there be copper in the metal, the glass would appear brick-red or dark brown when cold under the influence of the tinfoil.

The same pearl is played upon again with a good reduction flame for half a minute in order to reduce the copper to a metallic condition. If then, the glass appears transparent greenish or bottle green, there is iron present in the metal. If it does not become transparent, a longer blowing or better reduction flame is required.

BLOW-PIPE ASSAY.

Sec. 12. For the purpose of determining a silver mineral, it is not absolutely necessary to ascertain the amount of silver or the specific gravity. In some cases, however, it may be desirable to know both. To ascertain the silver by fire assay, would not answer; firstly, because there are generally different silver minerals in the ore, the separation of which in the required quantity would be almost impossible, and because the fire assay is not accurate enough for this purpose.

In analyzing silver ore it is important to examine it with the magnifying glass, and to select pieces which are of the same quality in regard to toughness, color, lustre, and fracture. A preliminary examination with the blow-pipe will inform us of the compound, and enable us to know whether the different pieces are substantially the same. Rich silver ore should be examined carefully with the aid of a magnifying glass, and the small particles of native silver adhering to the mineral removed by a pair of pincers. This precaution must be observed, likewise, in selecting pieces for the determination of specific gravity.

One grain of the silver mineràl is sufficient for the blowpipe assay, and, if properly conducted, will always yield the same amount of silver, as found in the classification, provided it is a mineral with an invariable amount of silver, for instance silverglance.

For the purpose of assaying, the mineral must be pulverized in the agate mortar to the finest powder. Tough silver ore can be used in little pieces. Of this prepared ore, 1 grain or $\frac{100}{1000}$ is weighed out on the assay balance (see Sec. 3) with the utmost precision, and emptied into the mixing scoop (see Fig. 3). Some powder always remains in the balance cup which must be swept with a little hair brush into the capsule. (Fig. 3.)

The required quantity of borax glass and lead (Sec. 4) depends on the ore. About $\frac{80}{1000}$ of borax will generally answer, but if the test appears difficult to be melted, some more borax may be added during the smelting operation. Three times as much lead as silver ore should be used. If there is a great deal of copper in the mineral, a double quantity of lead is required.

The ore, lead, and borax are mixed carefully in the capsule with an iron spatula. When this is done, the spatula is brushed above the capsule and the mixture introduced into the paper tube (Sec. 3). To effect this, the paper tube is gently held between the thumb and forefinger. The capsule is held with the thumb and forefinger of the right hand, and the mouth of the capsule introduced so far into the tube that it can be held, together with the tube, by the left hand.

The assay slides into the tube by gently knocking with the spatula upon the capsule. The remaining dust must be swept into the tube with the hair brush. The tube is then closed, folding the paper with the fingers, and is thus prepared for melting.

For this purpose, a hole is made in a sound piece of

charcoal with the little drill (Fig. 2) deep enough to receive the prepared tube, which is introduced. The hole must be always made across the grain of the coal near the edge. The reduction flame is now conducted so as to cover the tube by easy blowing. A strong blast would carry off the paper which was folded, and cause some loss. When the tube paper is burned and the assay commences to melt, a strong heat must be applied with a pure reduction flame which is chiefly turned upon the borax, not upon the lead button. The latter gains in size by the joining of other globules. When the melting has proceeded so far that only slag and the lead button are perceived, the blowing must be interrupted, in order to bring the bottom of the tube, which may be yet under the slag, to the top, by tapping the coal with the finger. The application of the reduction flame is continued till no particles of lead are seen on the slag. The charcoal must be inclined in different directions, so that the lead button may touch all sides of the slag, gathering up the small lead globules. When this is performed, the oxydation flame must be used, chiefly for the purpose of driving out the sulphur. The assay must be kept as far from the flame as possible.

The blast is directed right on the lead button. It will soon be perceived, that many little lead globules arise in the slag, but they do not contain silver. The lead oxyd, produced by the oxydation flame, is dissolved by the borax, and, in contact with the glowing coal, reduced to the metallic state. After a time the melting is interrupted, and when it is observed that the lead button turns black on cooling, the blowing must be continued for half a minute more. If the lead, when hardened, has a light lead-color it may be broken out, with a pair of pincers, and cleaned of slag on the anvil with the little hammer.

The button is now ready for **cupellation**. The bone-ash cupel (Fig. 4, *b*) is heated in the oxydation flame, and the lead button introduced. The object of this operation is the oxydation of the lead, which is thereby separated from silver. The lead button in the cupel is played upon with the oxydation flame, using a strong heat. As soon as the lead becomes bright and active the cupel is held a little further off from the flame and a moderate heat is applied, but not so low as to stop the action. The flame must be directed on the lead. The litharge accumulates behind the button, which, when reduced to the size as represented by *c* (Fig. 4), must be separated from the litharge, by using more heat on and around the button, holding at the same time the capul a little inclined, so that the globule may slide off on a clean spot. The cupel is held nearer to the flame, which must play around the button, in order to heat the cupel, by which the litharge as it is formed sinks into the mass, leaving finally a pure silver globule. When the last particles of litharge are emitted, the globule assumes the color of the rainbow, which indicates the finishing operation requiring a few seconds more blowing. If there was too much copper in the assay, the silver button will appear dark or black. An addition of lead of the size of a pin-head, and treated again with the blowpipe for a moment will render the button bright.

The silver button is taken from the cupel, laid edgeways on the anvil and the adhering boneash hammered off. It is then weighed on the balance and the per centage is found directly by noticing the weight.

If required to express the per centage in ounces per ton, it may be calculated in the following way: for instance, one hundred parts of silver ore assayed in the described way, yielded a button, weighing sixty-three. This mineral contains, then, sixty-three per cent. of silver, consequently twenty times so much per ton of 2,000 lbs., of which one pound contains 14·58 ounces Troy. The calculation accordingly is:

$$63 \times 20 = 1260 \times 14\cdot58 = 18{,}370\cdot80 \text{ ounces per ton.}$$

The silver button, when dissolved in nitric acid, leaves sometimes a black particle of gold, which is always insignificant and without influence in classification.

The application of too much heat in cupellation may cause a serious loss of silver. It is therefore necessary to keep the cupel sufficiently far off from the flame and allow the litharge to become stiff, as soon as it parted from the metal. But also in this case a certain loss must be suffered, which, as found by experience, must be added to the weight of the button.

PROCESSES OF SILVER AND GOLD EXTRACTION.

The following extract from Plattner's Table of Silver Losses by Cupellation will serve our purpose:

If the silver button weighed:	Ores containing 30 to 59 per cent. of copper, required 11 parts of lead to 1 part of ore.	Ores containing under 7 per cent. of copper, or free from it, require 5 parts of lead to 1 of ore.
	The loss of silver in cupellation will be:	
80		0·44
70	0·82	0·40
60	0·74	0·36
50	0·56	0·32
40	0·55	0·27
30	0·50	0·25
20	0·45	0·22
10	0·40	0·20
9	0·35	0·17
8	0·28	0·15
7	0·23	0·13
6	0·20	0·11
5	0·18	0·10
4	0·16	0·09

According to this table the real amount of silver in the supposed assay of the silver ore (which, in the absence of copper was mixed with 5 parts of lead) must be equal to $63 + 0·37 = 63·37$ per cent.

SPECIFIC GRAVITY.

SEC. 13. The determination of specific gravity, for the purpose of the classification of silver ore is not absolutely necessary, but in some cases it may be of service.

The specific gravity of a mineral is its weight, compared with the weight of an equal body of distilled or pure water.

It is necessary to know the absolute weight of the mineral, then that of an equal volume of water, and to divide the first by the last. The mineral by immersion loses exactly so much in weight as an equal bulk of water weighs. The loss of weight of a mineral by immersion expresses, therefore, the weight of an equal volume of water. If the absolute weight of a mineral is divided by its loss under water, the quotient will show the specific gravity.

For instance, a piece of native gold (Comstock) weighs............ 183·8
The same piece under water................................... 170·8

Difference .... 13·0

The difference, 13, is the loss of weight by immersion, and $\frac{183\cdot8}{13} = 14\cdot1$ specific gravity of gold.

The blowpipe-balances are provided for hydrostatic weighings, as represented by Fig. 8. On the little hook of the scale, a, is fastened a fine silk thread prepared with a sling (for the mineral, which may weigh two or three grains). The scales are set to balance, including the thread. The mineral is then suspended on the thread, weighed, and the weight noticed. A small tumbler, c, is placed under the scale, a, and the suspended mineral, b, immersed about one-fourth of an inch below the surface of the water. If any bubbles are perceived on the mineral, they must be removed by the aid of a small hair brush. It is weighed again now, and the loss by immersion found.

HARDNESS.

Sec. 14. In regard to gold and silver ores, the hardness is not very important, inasmuch as there is no considerable variation among these minerals. Chapman's arrangement, by which the hardness can be ascertained without the use of minerals, representing the scale of hardness, will sufficiently answer our purpose.

Hardness = 1·5 yields with difficulty to the nail.

Hardness = 2·5 does not yield to the nail; does not scratch a copper coin, but is easily scratched by it.

Hardness = 3·5 scratches a copper coin easily, but is scratched by it with difficulty.

CHAPTER II.

DESCRIPTION OF GOLD AND SILVER ORES.

A. GOLD ORES.

SEC. 15. Gold appears mostly in metallic condition, but never free from silver. It is found generally in the form of grains, scales, dust, also in the shape of leaves, threads or crystals. It is not ascertained but supposed, that a part of the gold in iron pyrites does not exist in metallic state, but combined with sulphur, or with arsenic in the arsenical pyrites. The gold is found in combination with the following metals:

1. *Silver*—In different proportions. The gold of Gold Hill lode, N. T., contains forty-seven to fifty per cent. of silver; that of the Comstock lode thirty to forty-five. Gila River and Australian gold three to five per cent., according to the amount of silver the gold appears more or less whitish. Sixty per cent. of silver renders the alloy white.

On charcoal, treated with the oxydation flame, it gives sometimes a bluish-white coating of antimony. With borax, played upon with reduction flame, a reaction of copper may be observed.

2. With *Tellurium.*—It contains gold 26, silver 14, tellurium 59, with traces of lead, copper, and antimony, hardness 1·5, gravity, 5·7 to 5·8, lustre metallic, color light gray.

In an open tube it emits white fumes, and gives a gray sublimate of tellurium. Directing the flame on the sublimate, it melts into transparent drops. The fumes have a peculiar sour odor. On charcoal it melts to a dark gray globule. Played upon with the oxydadation flame it gives a white coating, which disappears with a bluish-green color under the oxydation flame. Continued blowing yields a yellow, bright gold button.

3. With *Tellurium* and *Lead.*—Gold 9, tellurium 32, lead 54, with traces of copper, sulphur, and antimony; H. = 1·5, Gr. = 7–7·2, color dark lead-gray.

In an open tube it fumes, and yields a gray sublimate, the upper part of which, formed by antimonous acid, can be driven away by the flame. On charcoal it fumes and gives two coatings: a white one, which is volatile, consisting of tellurous and antimonous acids and sulphate of lead; the other coating is yellow, less volatile, and consists principally of oxyd of lead. Continued blowing leaves a small metallic button, showing gold color when cupeled.

4. With *Mercury* and *Silver.*—Gold 36, silver 5, mer-

cury 58. The gold is found also alloyed with molybdenum, platinum, and rhodium.

B. SILVER ORES.

SEC. 16. Silver is found mostly in combination with sulphur, also alloyed with other metals and substances. It appears often in metallic condition.

1. *Native Silver*—Is found crystalized, in threads or filaments. It often contains a small amount of antimony, arsenic, iron, gold, or copper. The native silver (one variety) of the Comstock lode, N. T., contains: Silver 60–85, gold 1·9, lead 8–30, copper 1–5, $H. = 2·7 -3$, $Gr. = 10·6–11·3$. Heated on charcoal it becomes covered with lead globules, disappearing again when red hot. It gives a yellow coating of lead, and further off a bluish-white of antimonous acid. It colors the borax glass green with the oxyd of copper.

a. COMBINATION WITH SULPHUR.

2. *Silver Glance* (Sulphuret of Silver).—Silver 87, sulphur 12·9, $H. = 2·5$, $Gr. = 6·9–7·2$, lustre metallic, color and streak blackish lead-gray, streak shining. It may be cut like lead. On charcoal, it melts into a dark blue globule, generally emitting metallic silver on the surface on cooling, especially if a small particle of borax glass is added, which dissolves impurities. It yields a silver globule when melted with soda.

3. *Stromeyerite* (Silver-copper Glance).—Silver 50–53, copper 31, sulphur 15, H. = 2·5, Gr. = 6·2, lustre metallic, color blackish lead-gray. In a closed tube gives sometimes a little sulphur sublimate, in an open tube sulphurous acid. On charcoal it fuses to a steel blue globule, emitting sometimes metallic silver on cooling. Melted with soda, it gives a copper button, which yields silver when refined. It occurs in the Heintzelman mine (Arizona).

a'. A variety of this ore, containing 40–43 per cent. of silver with a dull blackish-blue color, streak shining, can be cut, occurs in Arizona.

4. *Sternbergite* (Sulphuret of Silver and Iron).—Silver 30–33, iron 36, sulphur 30, H. = 1, Gr. = 4·2, metallic lustre, color pinchbeck-brown, streak black. In thin laminæ flexible, resembling graphite. In an open tube it gives out sulphurous acid. It melts to a globule on charcoal, emitting silver, and follows the magnet.

a'. A variety of this ore is found in the Gold Hill lode, N. T. It consists of silver 33·25, iron 34·05, H. = 2·8, Gr. = 5·2, color dull bluish-gray; the fracture has a metallic lustre and dark lead-gray color. The powder is blackish-brown. It is found in small fragments of indistinct cubic shape. On charcoal it melts with a spongy appearance to a dull gray globule, following the magnet. A slight yellow coating indicates a trace of lead. In melting it gives out a great deal of sulphurous acid. Treated with soda a silver globule is easily obtained.

b. **Combinations with Sulphur and Antimony, or Arsenic.**

5. *Brittle Silver Ore.*—Silver 70, antimony 13·9, sulphur 15·7, H. = 2·5, Gr. = 6·2, lustre metallic, color and streak iron black or blackish lead-gray. In a close tube it decrepitates, melts to a globule and gives a blackish sublimate which turns red-brown when cold, consisting of sulphide of antimony. In an open tube it melts, evolving sulphurous acid, and fumes. On charcoal it fuses, and coats the coal white with antimonous acid. By continual blast, the coating assumes a pink color, derived from the oxyd of silver. It occurs frequently in the Comstock lode.

6. *Polybasite* (Eugen Glance).—Silver 64–72, copper 3–10, sulphur 17, H. = 2·5, Gr. = 6·2. It contains also antimony, arsenic, iron, and sometimes zinc. Lustre metallic, color iron black, streak black. In a closed tube it yields nothing volatile. In an open tube it gives antimonial fumes and sulphurous acid. It occurs also in Gold Hill lode, N. T.

a'. The polybasite of the Comstock lode contains 64 per cent. of silver. It gives in a closed tube, with the aid of the blow-pipe, a reddish brown sublimate with a yellow edge. In an open tube, white fumes arise and some white sublimate deposits. On charcoal, with the reduction flame, it evolves an odor of garlic. Played upon with the oxydation flame, it gives out sulphurous acid and a white coating of antimonous acid. It melts to a globule with a metallic lustre. If the hot blast is changed suddenly to a cold one, and directed on the globule, holding the blow-pipe point close to it, metallic silver is emitted. If the cold blast is stopped too soon, the silver will disappear again.

7. *Miargyrite.*—Silver 35·8, antimony 42·8, sulphur 21, H. = 2·5, Gr. = 5·2–5·4. Lustre metallic adamantine, color iron black, streak dark cherry red. In a closed tube it decrepitates, melts easily, and gives out a sublimate of sulphide of antimony. In an open tube sulphurous acid and antimonial fumes are emitted, depositing a white sublimate of antimonous acid. On charcoal it melts quietly, emitting sulphurous acid and antimonial fumes. It covers the coal with a white coating, which becomes pink colored by continual blast. Melted with soda a silver button is obtained, which, treated with borax and tin, reacts on copper.

8. *Dark Red Silver Ore* (Pyrargyrite, Antimonial Blend). — Silver 58·9, antimony 23·4, sulphur 17·5, H. = 2·5, Gr. = 5·7. Lustre metallic-like adamantine, color dark red, powder cochineal-red. In a closed tube, by the aid of the blowpipe, it yields a sublimate of sulphide of antimony, black while hot, but varying from red to reddish-yellow when cold. In an open tube it gives antimonial fumes and sulphurous acid. On charcoal it melts easily and deposits a white coating of antimonous acid. With soda it gives a silver globule. It occurs also in the Gold Hill lode, N. T.

9. *Light Red Silver Ore* (Proustite, Arsenical Blend). Silver 65·4, arsenic 15·1, sulphur 19·4, H. = 2·5, Gr. = 5·5–5·6, color similar to dark red silver ore, but lighter. Behaves like the preceding, except the arsenical fumes.

10. *Xanthocone.*—Silver 64, arsenic 13·4, sulphur 21·3, H. = 2, Gr. = 5–5·2, color dull red to clove brown, powder yellow. When heated in a closed tube it becomes dark red, melts and gives some sublimate of sulphide of arsenic. While hot, it is dark brownish red, and red to reddish-yellow when cold. In an open tube and on charcoal it behaves like the preceding.

11. *Silver Fahlerz* (Argentiferous gray copper ore). Silver 17·71–31·29, antimony 26·63–24·63, sulphur 23·52–21·17, copper 25·23–14·81, iron 3·72–5·98, zinc 3·10–0·99, lustre metallic, color light steel gray. In a closed tube it sometimes decrepitates, melts and gives, by aid of the blowpipe, a dark red sublimate of tersulphide of antimony with antimonous acid. In an open tube it fuses, gives antimonial fumes and sulphurous acid. On charcoal it fuses easily and gives a bluish-white coating of antimonous acid and antimonial fumes. There is also a yellowish coating close to the test which appears white on cooling. This coating is created by oxyd of zinc.

a'. The Reese River ore from the Comet lode seems to be a metamorphosed silver fahlerz. The sulphur is represented by carbonic acid, so that almost all copper and silver is a carbonate. It contains silver 22·35, copper 17, antimony, and some lead. It has a dull greenish-black or black color, streak shining, powder greenish-gray. In a closed tube it yelds nothing volatile. In an open tube some sulphurous acid can be observed. On charcoal fuses slowly, but boils up suddenly in contact with glowing coal, leaving a button of silver and copper. This button, when played upon with the oxydation flame on another spot of the charcoal, gives first a bluish coating of antimonous acid, then a yellow one nearer to the assay of the oxyd of lead. The silver can be separated from copper by cupellation with lead.

b'. The silver fahlerz of Sheba lode (Humboldt) contains silver, 8·20, gold 0·008, some antimony and lead, but very little copper. It has a light-gray metallic lustre. It is also called gray silver ore.

c. COMBINATION WITH CHLORINE, BROMINE, AND IODINE.

12. *Horn Silver* (Chloride of Silver).—Silver 75·2, chlorine 24·6, H = 1·5, Gr. = 5·5–5·6, lustre adamantine, color gray, greenish or blackish, streak shining. It looks like horn or wax. It is translucent and may be cut like wax. Occurs frequently in the Comstock and Gold Hill lodes; also in California. It fuses in a candle flame. On charcoal it is easily reduced and gives an odor of chlorine. If treated under the reduction flame with an addition of copper, it forms a chloride of copper and colors the flame azure-blue.

13. *Embolite* (Chlorobromide of Silver).—Silver 66·9 to 75, H = 1–1·5, Gr. = 5·3–5·4, lustre resinous, color yellowish-green or green. On charcoal it fuses easily, evolves vapors of bromine, and gives metallic silver. Mixed with oxyd of copper it colors the flame greenish-blue.

14. *Bromyrite* (Bromic Silver).—Silver 57·56, bromine 42·44, H. = 1–1·5, Gr. = 5·8–5·6. In a closed tube, treated with bisulphate of potassa, it emits brown vapors. On charcoal it fuses easily and yields a globule of silver. It is yellow or greenish, and may be cut like chloride of silver.

15. *Iodyrite* (Iodid of silver).—Silver 46, iodine 54, H. = 1·5, Gr. = 5·5. Lustre adamantine. Color yellow, also greenish. It is translucent. In scale shape it is always lemon-yellow. When heated in a closed tube it becomes fire red, but assumes its former color when cold. It fuses easily, and gives, by the aid of the blowpipe, a reddish yellow sublimate, getting lemon-yellow on cooling. With bisulphate of potassa, it emits beautiful violet vapors. In an open tube it gives an orange sublimate, lemon-yellow on cooling. On charcoal it assumes a fire red color before it fuses, and spreads on the coal and yields many minute silver globules. With an addition of oxyd of copper, it makes an intensely green flame with a bluish tinge.

16. *Iodid of Silver and Mercury.*—Silver 40–42, iodine, quicksilver, and sulphur (chlorine?) Color dull, dark red. Streak shining. Powder dark red, but changes soon into lead-gray if exposed to the light. In a closed tube it gives three sublimates, separated in rings. The nearest to the assay, is black (sulphide of mercury), the second yellow, (subchloride of mercury?) the third is gray (metallic mercury). An addition of bisulphate of potassa causes it to yield violet vapors, which come from the iodine. In an open tube it gives the same sublimate, but the black is very slight; it gives also yellow fumes. A gold particle in the tube becomes amalgamated. Litmus paper at the upper end is colored red by the sulphur. Heated on charcoal, it turns black, fuses easily, and yields silver globules. Melted with soda, it draws partly into the coal. If this crust is broken out, and laid on a blank piece of silver, with a drop of water, the sulphur in it will cause a black spot on the silver. Heated with a small piece of pure lead, it gives a beautiful green coating with a yellow border nearest the assay. This coating (iodine and lead) is far off from the test. With copper oxyd, like the iodyrite.

This mineral occurs to my knowledge only in the Heintzelman mine *(Arizona)*.

d. COMBINATION WITH ANTIMONY.

17. *Antimonial Silver.*—Silver 77–84, antimony 23–16, $H. = 3.5$, $Gr. = 9.4$–9.8. Lustre, metallic; color and streak, silver-white. On charcoal it fuses easily to a

globule, coating the coal white. A continual blast renders the white coating reddish.

e. Combination with Selenium.

18. *Naumannite* (Selenid of silver).—Silver 73, Selenium 26, H. = 2·5, Gr. = 8. Lustre, metallic; color, iron black. It melts easily on charcoal, but with intumescence in the reduction flame. It emits the selenium odor of rotten radish. With soda it yields metallic silver.

19. *Eucairite* (Selenid of silver and copper).—Silver 43·1, selenium 31·6, copper 25·3. Lustre, metallic; color, lead-gray. On charcoal it melts to a gray metallic globule, fumes and reacts on borax with copper. This mineral is soft, and can be cut with a knife.

f. With Tellurium.

20. *Hessite* (Tellurid of silver).—Silver 62·42, tellurium 36·96, iron 0·24, Gr. = 8·4–8·6. Lustre, metallic; color, lead-gray, or steel-gray. It is soft, and can be cut like lead. According to Mr. Blake, this mineral is found in California also. He describes the reaction as follows :

"In an open tube the mineral fuses quietly, coloring the glass a bright yellow under assay ; a white or gray sublimate is deposited at a short distance, immediately over it, which, on being heated, fuses into transparent drops resembling oil. On charcoal it fuses to a leaden colored globule, which on cooling becomes covered with dendrites. This globule flattens under the hammer. With the addition of soda, a silver globule is obtained."

f. WITH BISMUTH.

21. *Bismuth Silver.*—Silver 60, bismuth 10, copper 7·8, and some arsenic. Lustre, metallic; color, tin white or grayish. On charcoal it melts easily, covering the coal dark orange. It is yellow while hot, and lemon-yellow when cold. The oxyd of copper in it colors the borax green when melted on charcoal.

g. WITH MERCURY.

22. *Silver Amalgam.*—Silver 34·8–26·2, quicksilver 65·2–73.7, H. = 3·5, Gr. = 13.7–14. Lustre, metallic; color, silver-white; brittle. In a closed tube the mercury sublimates.

23. *Arquerite* — Silver 86·49, quicksilver 13·51. It behaves like the amalgam.

CHAPTER III.

METALLURGY OF GOLD AND SILVER ORES.

FIRE ASSAY OF ORES.

The modes of assaying here described are those best suited for the purpose of the miner and millman. The apparatus and method of procedure are as simple as is possible, consistent with correctness.

TOOLS.

SEC. 17. A fine assay balance, as described in Sec. 3.

A pair of less delicate scales, capable of weighing about three ounces. The weights are Troy ounces; one ounce divided into $\frac{100}{100}$. Such a scale may cost in San Francisco from $10 to $12.

French clay crucibles No. 7; some glass matrasses; dry cups (small crucibles of pipe clay); a fine wire cloth sieve on a wooden frame, 50 holes to the inch (2,500 to a square inch); one pair of crucible tongs; one pair cupel tongs; a pair of pincers; a tooth brush and an iron mortar; two or three muffles, 10 inches long, 4 inches wide, and 3 inches high.

Cupels.—The fabrication of cupels requires a brass mould and boneashes. The bones are burnt white for this purpose. Pieces not well burnt, showing a black inside, do not answer. The bones are then pulverized and sifted through a fine sieve, forty holes to the inch. The bone powder is sprinkled with water, mixed and rubbed with the hands till it appears uniformly moistened and allows itself to be formed into a ball by a squeeze of the hand, without wetting it. The mould is then filled and beaten with the pestle by a wooden mallet. The pestle is drawn out in a twisting way and the cupel is pressed out by the ball of the hand.

MATERIALS.

Sec. 18. 1. *Litharge.*—Sifted, well mixed, and kept always under cover. One ounce and a half, mixed with 10 grains of wheat flower, melted in a crucible, will yield a button, which must be cupeled, and the weight of the small silver grain noticed. This weight must be subtracted from all assays, where this quantity of litharge was used.

2. *Wheat Flour.*—It is used instead of charcoal to reduce a part of the litharge to lead. A mixture of one hundred parts of soda and twelve parts of wheat flour serves as a flux for lead assays.

3. *Soda* (Carbonate of Soda).—If this soda is crystallized, it must be exposed for some time to the air, till

it turns into a white powder, thus losing half of its water of crystallization. Bicarbonate of soda or soda-ash answers the purpose likewise.

4. *Glass.*—Broken glass is pulverized in an iron mortar and sifted. It serves as a flux.

5. *Salt.*—Common table salt is freed from water of crystallization by heating in a sheet iron box, like a coffee roaster, till the decrepitation ceases. The salt fuses quicker than the assay mixture, and prevents the contact between air and the assay.

6. *Iron.*—It is used in small pieces of wire one-fourth or three-sixteenths inch thick, cut into pieces of different lengths from one-fourth to one-half inch. The purpose is the desulphuration of the sulphurets.

GOLD AND SILVER ASSAY.

SEC. 19. The ore intended for the assay must be broken first in small pieces, of which about one pound is taken without selection, afterwards well mixed and pounded finer. From this another portion is taken, about three ounces, pulverized and sifted through the sieve, described in Sec. 17. Ores from new lodes should be assayed for the purpose of ascertaining the highest yield, and the darkest parts of the ore selected and pulverized. In taking samples of tailings, it is the safest way to procure three or four boxes of about one cubic foot each, and have them filled from the discharge at different times, say at the commencement, at the middle and at the end of the discharge time. The box must be removed as soon as the water reaches the brim. After some hours, when the sand and mud have settled, the water is poured off, and the contents of all the boxes mixed and dried. These tailings are spread out, and small quantities are taken from different parts of the surface to the amount of three or four pounds. This sample is mixed well and treated in the same way, taking four or five ounces, which must be sifted and pounded, if any coarse sand is present. The mortar must be cleaned carefully after pounding each sample, especially if it was rich. In this latter case the mortar must be washed, or a small piece of quartz pounded in it and then wiped out with a clean cloth.

The prepared sample and fluxes are weighed in the following proportion:

a. ORES OR TAILINGS CONTAINING BUT LITTLE SULPHURETS.

Ore	250 grains.
Glass	125 grains.
Flour	8 grains.
Litharge	1½ ounces.
Soda	1 ounce.

b. ORES CONTAINING ABOUT FIFTY PER CENT. OF SULPHURETS.

Ore	250 grains.
Glass	125 grains.
Iron	50 grains.
Litharge	1½ ounces.
Soda	1 ounce.

c. Ore being nearly all Sulphurets.

The mixture is like the preceding, but double the amount of iron (one hundred grains) must be used.

The soda and litharge are first introduced into the crucible, then the balance of the mixture. By aid of an iron spatula, the end of which is rounded, ore and fluxes are mixed in the crucible carefully over a sheet of paper. If in mixing, some powder should be spilled, the test can not be considered proper. It is better to repeat the weighing. If there are several assays made at the same time, the crucibles must be numbered outside with red chalk. The charged crucible is tapped several times against the table, in order to settle the mixture, and is covered with salt, about one-fourth of an inch deep.

Thus prepared, the assay is ready to be melted. The crucible is placed in the middle of the furnace, on the muffle, or, if there are two or three assays to be melted, the crucibles, standing on the muffle must touch each other, leaving space between them and the walls, as represented in Fig. 9.

The furnace is charged with charcoal only to the top of the crucibles, and set on fire by some live coal. It is not advisable to fill the furnace entirely, unless No. 8 crucibles are used, because the assay effervesces in melting and may overflow. The covers are taken off, the charcoal, by replacing, kept level with the top of the crucibles until the melting mass has gone down. The assays are covered again, the furnace filled with coal to the line, *b*, and closed by an iron cover, *a*. The cupels are placed now in the muffle, *d*, and closed by the shutter, *e*.

In about half an hour the charcoal is burnt down so far that the tops of the crucibles are exposed. By means of the crucible tongs the covers are removed first, the crucibles then, one after another, are seized with a pair of tongs (as represented in Fig. 10 *c'*) and the contents poured into an iron mould (Fig. 12), containing four or five hollows, three-fourths of an inch deep. In pouring, the crucible must be kept close to the mould and slowly inclined, by degrees almost perpendicularly, so that the slag may drop out entirely. This melting operation takes nearly one hour's time. Care must be taken not to let the coal burn down below the line, *f*, else the upper part of the crucible might get too cool.

The lead button in the mould is cooled off in one or two minutes, freed from slag by hammering it into a square shape, and, by means of the cupel tongs, introduced into the light red hot cupels. It melts in a short time, and the oxyd on the surface draws to the sides of the cupel. The lead appears bright, and fuming, and minute spots of litharge, constantly appearing, glide to the sides, being absorbed by the cupel mass. The muffle is kept open, but the ash holes, *g* and *g'*, are closed. The front edge of the cupels will cool off by the draft, and appear dark red; the button inside is bright. The temperature must be kept as low as possible, but sufficient to keep the button in action. If, however, the temperature be too low, and the lead, covered with oxyd, appearing dull, a piece of live charcoal is placed in front of the cupels, and the muffle is closed. By increasing the heat, the button soon resumes its activity, the muffle is opened again, and the coal removed. The furnace must be charged with charcoal, so that the muffle is always covered.

The button becomes gradually smaller, but looks bright as long as litharge is separating from it. As soon as the last particles disappear, a play of rainbow colors is perceived on the remaining silver button, indicating the end of the cupellation.

Generally, the crucibles are taken out of the furnace, when all the coal is burned down, placed in a cool place, and broken when cold. It is, however, a great saving in time, if the melted mass is discharged in the described way. A comparative trial will show that the results are equal, if well performed.

If there is not too much importance connected with the assay, the crucibles may be used over four or five times, but care must be taken not to use crucibles in which rich assays were made.

The silver button is freed from adhering to boneash by hammering edgeways and weighed. The weight, multiplied by 1·16, gives the amount of ounces per ton of ore of 2,000 lbs., which may be illustrated by an example. For instance, a button is found to weigh—

300
50
6
———
356 × 1·16 = 412·9 ounces per ton of ore.

After weighing, the silver button is introduced into a glass tube, adding about half an ounce of pure nitric acid, and heated by the alcohol flame. It soon begins to boil, emits reddish-brown vapors and leaves the gold, if any in the assay, in undissolved particles of a black color in the tube. The nitric acid, containing the dissolved silver is poured off slowly, and the tube filled with distilled water. When all the particles of gold have settled, this water must be poured off again carefully and the tube filled once more with water, to the brim.

The tube is then covered with the dry cup and quickly turned over, as represented in Fig. 11. The gold falls to the bottom of the cup, but, being very light and sometimes in minute particles, the tube must be kept for a while in this position, till no suspended particle is visible. It requires now some practice to lift the tube without spilling some water, which would invariably carry out some gold. The easiest and surest way is to gradually lift up the tube, till the water, the brim of the tube and the dry cup are level (Fig. 11, a). A slide of the tube in the direction of a' leaves the gold in the cup undisturbed. A slight tapping of the cup will bring the gold particles together, the water is poured off and the cup dried over the alcohol flame till the gold assumes a yellow color.

This gold is now carefully weighed and calculated upon as follows:

The gold was found to weigh, for instance, $\frac{35}{1000}$, and the silver button before dissolving $\frac{356}{1000}$. If the gold is subtracted from the silver which contained this gold, we find thus the pure silver—

$$356-35 = 321 \text{ silver} \times 1\cdot16 = 372\cdot3 \text{ ounces per ton.}$$
$$\text{and } 35 \text{ gold } \times 1\cdot16 = 40\cdot6 \text{ " "}$$

To find the value, the ounces of gold must be multiplied with 20·67 and those of silver with 1·30. These numbers in their fractions are not perfectly correct, but will serve our purpose. Continuing the calculation we find—

$$\text{Silver} = 372\cdot3 \text{ ounces} \times 1\cdot30 = \$483\cdot99$$
$$\text{Gold} = 40\cdot6 \text{ " } \times 20\cdot67 = 839\cdot20$$

Total value................$1,323·19 per ton.

In assaying gold ore, the button will not dissolve in nitric acid. In this case it must be melted (after weighing) on a piece of charcoal before the blowpipe, with the addition of three times its weight, of pure silver and then dissolved and treated as above described.

In case the ore for the assay has been weighed out by half an ounce, equal to two hundred and forty grains, the calculation is made in the same way as before with the exception, that the number 1·215 must be substituted for 1·16. The procedure of the preceding example would be as follows:

The weight of the button was three hundred and twenty-one. This multiplied with 1·215 will give the amount of ounces per ton of ore of 2,000 pounds.

32 (321) × 1·215 = 390 ounces. The quantity of fluxes used for two hundred and fifty grains of ore will also serve for half-ounce assays.

LEAD ASSAY.

SEC. 20. In using lead ore for the purpose of melting silver ores, the amount of lead in the ore must be ascertained. The lead ores of Nevada Territory and California are represented chiefly by sulphuret of lead (galena), but also to some extent by carbonate of lead.

The lead ore must be sifted in the same way as the silver ore (Sec. 19) and mixed with fluxes in the following proportions:

a. ORE CONTAINING SULPHURET OF LEAD.

Ore..	½ ounce.
Soda with 12 per cent. of wheat flour..........	1½ ounces.
Glass.......................................	½ ounce.
Iron..	100 grains.

b. ORE CONTAINING CARBONATE, OXYD, OR MOLYBDATE OF LEAD.

The mixture is the same as that of the ore, *a*, but no iron is taken, as there is no sulphur in the ore. If it is pure galena, one hundred grains of iron are used. If therefore the ore, *a*, contains a great deal of earthy matter and less galena, or if the ore, *b*, is mixed with sulphurets, the quantity of iron may be taken more or less, according to the supposed quantity of sulphurets. The melting is performed in the same way as is done with the silver assay. The lead button, resulting from the assay, is weighed on the ounce scale. As one ounce is divided $\frac{100}{100}$, the per centage of lead in the ore is easily found by doubling the weight of the button.

CHAPTER IV.

EXTRACTION OF GOLD.

SEC. 21. The process of extracting free gold, and the manipulation itself, is very simple, requiring only a proper friction and contact with quicksilver. But there are combinations of gold with other substances in California, refusing to liberate the gold by friction. Such ore, as arsenical, and some iron pyrites, or tellurium of gold, require a different treatment.

There are two principal methods of gold extraction: By amalgamation, and by chlorination.

A. BY AMALGAMATION.

a. AMALGAMATION IN THE BATTERY.

For this purpose the batteries are provided with amalgamated copper plates, three to five inches wide, having the length of the battery; one at the discharge, the other at the feed side; the latter being protected by the iron feed plate. They are fixed with a pitch of about thirty-five to forty degrees towards the dies. Other batteries are so constructed as to have sufficient space where the amalgam may accumulate. In this case, the stamps are three to four inches apart from one another and from the sides of the mortar; also, iron vertical grates inside the sieves are in use. The amalgam deposits readily between the rods. The amalgam adheres best to copper plates which are coated with quicksilver. This is performed by rubbing quicksilver on the copper with a piece of cloth tied to a wooden handle, using some drops of nitric acid, which may be diluted with the fourth part water.

The quantity of quicksilver depends upon the quantity of gold in the ore. One ounce of gold requires one ounce of quicksilver, but when the gold is very fine, one and one-fourth to one and one-half ounces may be used. The quicksilver is introduced every half-hour, or every hour by the feeder, during the stamping, in each battery, in portions of one-quarter of an ounce, more or less, as the ore requires. This may be observed at the discharge. When the amalgam appears very hard or dry, some more quicksilver may be used; but if, on the contrary, the amalgam is too soft, or if quicksilver drops are perceived, less quicksilver must be introduced.

The amalgamation goes on very rapidly. One hour after the quicksilver is put in, no yellow gold particles come out of the battery, except in cases when the quartz, containing lead, antimony, or other volatile metals, is burned for the purpose of rendering it easier to break. Many particles of gold appear coated, and are discharged without being amalgamated.

If the proper proportion of quicksilver, and the regular times of charging be observed, when the ore contains heavy gold (800 fine), sixty to seventy-five per cent. may be saved in the battery and the copper-plated platform; but light gold (300 to 400 fine), like Washoe gold, gives a less favorable result. A great many fine particles of amalgam adhere together, involving also manganese scum, if present, and form small spongy, blackish lumps, which are so light as to float, and on account of being coated with foreign matters, will not unite with the accumulated amalgam. Of this amalgam but very little can be saved; it floats over blankets, copper plates, or ripples.

It is therefore an error to use quicksilver in the battery, if concentration is in use, and the tailings are not saved. The finest gold is easier retained by concentration than this floating amalgam. There is also no evidence of any advantage in battery amalgamation, when the whole mass of pulverized rock is amalgamated in pans, unless the mass or the concentrated part is intended for roasting.

b. AMALGAMATION ON COPPER-PLATED PLATFORMS, TROUGHS, AND OTHER COPPER FIXINGS.

(Very imperfect, and mostly abandoned.)

c. AMALGAMATION IN ARRASTRAS.

This is a very old, primitive method, but gives comparatively a good result on the free gold, if, under good management, sufficient time is allowed. The construction is well known. There is a stone bottom, ten to fourteen feet diameter, and wooden sides, twenty to twenty-five inches high. Four or six large stones are dragged in a circular way by chains, fastened to four arms of the upright shaft. They make from six to ten revolutions per minute, and grind one and one-half to two tons of rock (broken in pieces as large as a hen's egg or smaller) in twenty-four hours. This is, however, too much for a proper amalgamation.

When in motion, the arrastra is charged with about two hundred pounds of ore, with some water. One quarter of an hour afterwards the balance of the whole charge, from four hundred to five hundred pounds, is introduced. As soon as the ore is turned into mud, one or two ounces of quicksilver are pressed through a dry cloth over the thick pulp. A sample is taken from time to time with the horn spoon, washed and examined. When free gold is perceived, after the amalgamation has been going on for some time, some more quicksilver may be added. The first charges require a little more quicksilver. After four or five hours the pulp is diluted with water and discharged. The next charge is treated in the same way, and so on till one hundred or one hundred and fifty tons are worked through. The quicksilver must be used always in proportion with the gold, one or one and one-half ounces to an ounce of gold. The amalgam imbeds in the crevices of the bottom, and must be always dry. The use of too much quicksilver makes the amalgam thin, causes an imperfect amalgamation,

PROCESSES OF SILVER AND GOLD EXTRACTION. 63

and a loss in quicksilver, which is often found beneath the bottom rock.

d. AMALGAMATION IN IRON PANS.

The pan amalgamation is a highly improved arrastra amalgamation, and at present the most perfect gold manipulation. The two conditions, friction and contact with quicksilver, are accomplished in a high degree by Wheeler's pans, the description of which will be found in Section 50. The supposition that a slow motion is favorable for the amalgamation is erroneous and entirely refuted by recent experience. To what degree, however, velocity may be advantageously increased is not yet ascertained; but sixty revolutions per minute of a properly-constructed muller answers most satisfactorily, but the quicksilver is destroyed by friction to some degree.

There is no chemical process required for amalgamation of gold, except with such ore as is mentioned in Sec. 21. By the pan manipulation the gold is extracted as close as ninety-five per cent. of the fire assay. The loss of gold in the pans does not result from defective amalgamation, but from improper discharge.

Ores, containing gold in such condition, that it cannot be liberated by grinding, must be subjected to roasting without salt, before treating in pans.

The treatment of gold ores does not differ from that of silver ores, except that no heat and no chemicals are required.

B. BY CHLORINATION.

SEC. 22. This process is based on the property of chlorine, which enables it when placed in contact with gold to form a terchloride of gold without the application of heat. The silver, when in the metallic state or as sulphate, undergoes the same change, forming chloride of silver, but the chloride of gold is soluble in water, chloride of silver only in a hot solution of salt.

This process is executed in Nevada, California. Another establishment, belonging to Mr. Deetken, in San Francisco, beneficiates concentrated sulphurets from different parts of California.

The chlorination of gold ores is a very simple process, still there are some delicate points in it. Comparatively, very few hands are employed; and there is neither motive power nor steam. This process, if well managed, extracts the gold very closely. Coarse gold particles, generally not found in the tailings, would resist chlorination, or require too much time. According to Mr. Deetken's experience, low gold (in fineness) in the tailings is preferable, it being sooner transformed into chloride.

The tailings are subjected first to calcination in a roasting furnace, without being sifted. No salt is used, as it sometimes causes a loss of gold. The roasting is performed in the usual way by stirring the mass at a low temperature till all the sulphurets or arseniurets are decomposed. An addition of charcoal powder favors the roasting. After six or eight hours, when no odor of sulphurous acid is observed, the ore is discharged, spread on a proper place and cooled. The tailings or ore is then sprinkled with water and shoveled over several times. A little too dry or too wet has a great influence on the result of chlorination.

When moistened, the stuff is introduced into wooden tubs about seven feet in diameter and twenty-five or thirty inches deep. These tubs have a prepared bottom, which allows the entrance of chlorine gas from beneath into the mass of tailings. Near the bottom are two holes, one for the discharge of the solution, the other communicates by a lead pipe with a leaden gas generator. The generator is filled to a certain height with peroxyd of manganese and salt. Sulphuric acid is introduced by a lead pipe. As soon as the mixture becomes hot, by the fire underneath the generator, the chlorine gas commences to be evolved and enters the tub through the connecting lead pipe.

After some hours the whole mass is strongly penetrated and the greenish gas lies heavy on the tailings. The tub is closed by a wooden cover. In this condition it remains for ten or fifteen hours, when the cover is removed and clean water introduced. As soon as the water reaches the surface of the tailings, the discharge pipe is opened, and the water, containing the dissolved chloride of gold, is led into glass vessels. An addition of sulphate of iron, precipitates the gold in metallic condition as a black-brown powder. If there are silver sulphurets in the ore, they, by roasting without salt, are converted mostly into sulphates, and in subsequent contact with chlorine, into chlorides which are not soluble in water, and remain in the tailings. The gold is therefore 995 fine.

CHAPTER V.

EXTRACTION OF SILVER.

The extraction of silver, as practiced in Nevada Territory and California, may be described as follows:

I. WET PROCESS.
> AMALGAMATION IN IRON PANS.

II. ROASTING PROCESS.
> *a.* AMALGAMATION IN BARRELS.
> *b.* AMALGAMATION IN VEATCH TUBS.
> *c.* AMALGAMATION IN IRON PANS.

III. COLD PROCESS.
> AMALGAMATION IN HEAPS (PATIO).

IV. MELTING PROCESS.
> EXTRACTION OF SILVER BY LEAD.

I. WET PROCESS.

SEC. 23. About two years ago, Mr. Smith attracted attention by his "Smith's Process," using the ore without roasting, in iron pans four feet in diameter. The surface of the bottom was diminished by a center-piece and by many shoes, so that only fifty lbs. of ore could be charged, and worked for five to six hours. Since that time there has been little improvement in the exclusively chemical part of this process; but the whole pan arrangement has been gradually so perfected, that now-a-days a four-foot pan (Wheeler's) is charged with seven hundred and fifty or eight hundred pounds of ore, and the amalgamation is finished in three hours including charge and discharge. But this mechanical improvement comprehends also that of the chemical part. It is known that friction and iron decompose tough silver sulphurets, without chemicals. Friction and iron are powerful chemicals in themselves. Silver ore, treated with chemicals in a stone arrastra for twelve hours, will not yield half so much silver as one Wheeler's pan in three hours, without any chemicals.

A great advantage of this process, is the working of unroasted ore. Dry crushing, injurious to machinery, is not required; the immediate working of the pulverized ore prevents the waste, and the silver, resulting from this process, is generally very fine (970 to 997). It is comparatively a cheap process. But, although improved, the result of this method cannot be considered yet quite satisfactory. In regard to chemicals, an important discovery can hardly be expected. Almost everything, between blue vitriol and tobacco or tea decoction, which reasonably or unreasonably permitted a supposition that it might effect the decomposition of the sulphurets, has been experimented on, and yet, there are no chemicals known by means of which more than fifty or sixty per cent. of the silver can be extracted. If a higher per centage is obtained, it is on account of the gold, or the prevalence of silverglance.

But the decomposition of silver sulphurets does not depend on chemicals alone, as is demonstrated by Wheeler's pans, by which the silver can be extracted from ten to fifteen per cent. closer than in common pans. The result, however, depends very much on the quality of the ore, if the latter is not roasted. It must be observed that the above per centage has no reference to gold.

Not all silver combinations are suitable for the wet process. Sufficient experiments, however, have not yet been made as to how the different silver ores behave in the pans, but it seems that the difficulty of decomposition grows with the amount of sulphur in the ore, and especially with that of antimony. Arsenical combinations are more easily worked than antimonial. Pyrites of arsenic, iron, and copper, are not affected at all by chemicals. In cases where such pyrites are argentiferous, or antimonial combinations are prevalent, the roasting process must be adopted.

CHEMICALS USED IN DIFFERENT MILLS.

SEC. 24. 1. *Sulphate of Copper or Blue Vitriol.*—It consists of 31·72 oxyd of copper, 32·14 sulphuric acid,

36·14 water. It dissolves easily in water. In contact with iron, metallic copper is precipitated, and in the presence of quicksilver amalgamated. The sulphuric acid combines with iron oxyd to form sulphate of iron. If there are no sulphurets in the ore, or if in proportion too much blue vitriol is used, the amalgam retains almost all copper, which is precipitated with iron; but in contact with silver sulphurets, under some circumstances of galvanic chemical action, the copper is expelled again from the amalgam, and enters probably into combination with sulphur. As soon as the blue vitriol is introduced, it may be observed that a great part of the bottom gets instantly amalgamated. The surface, to which the quicksilver adheres, is not iron, but copper, being precipitated by iron. This copper amalgam is removed by friction of the muller, and is taken up by the quicksilver; and yet, the metal after retorting and melting shows only a trace of copper, if the right proportion of the blue vitriol and silver sulphurets was observed.

In many cases the amalgam after retorting looks black, yielding a good deal of matt in melting. This matt appears tough, is dark grayish-blue in color, and contains silver in different proportions up to eighty per cent. and sometimes a considerable amount of copper. The silver bar, however, is over 970 fine (*). The matt,

---

* The fineness must be always understood in relation to 1000 parts of the metal; 970 fine means 970-1000 pure silver and gold, and 30-100 base metals.

when it has a yellowish color, contains mostly sulphide of iron.

The blue vitriol must be used always in solution. In this state it is mixed sooner with the pulp, and kept in suspension, having thus better opportunity to act immediately.

2. *Sulphate of Iron, Copperas, or Green Vitriol.*—It consists of 27·19 protoxyd of iron, 31·02 sulphuric acid, and 41·79 water. Exposed to the air, it turns into white powder. This salt is obtained by dissolving iron in diluted sulphuric acid.

3. *Bisulphate of Soda.*—It is composed of 63 sulphuric acid, 37 soda. This salt is obtained in the acid factories producing nitric acid, from Chile saltpetre. Its operation in the pan is due to the fact that it readily parts with one portion of its sulphuric acid, leaving sulphate of soda.

4. *Alum.*—Potassic alum consists of 33·76 sulphuric acid; 10·82 alumina (clay); 9·59 potassa; 45·47 water. The sodic alum contains 34·94 of sulphuric acid.

5. *Sulphuric Acid.*—This chemical seems to act partly by direct decomposition of silver sulphurets. Instantly after being introduced into the ore, it emits sulphuretted hydrogen. The silver may be oxydized by the oxygen which is disengaged by the parting hydrogen and converted by the acid into sulphate, or it may be set free in

metallic condition. In both cases it can be amalgamated without salt. But the salt, under the action of the sulphuric acid, creates muriatic acid. This, however, may be limited. A considerable part of the sulphuric acid is engaged in dissolving iron, forming sulphate of iron, setting hydrogen gas free. There is always so much iron in the pan, from the wear of the stamps when the ore is crushed, and from the shoes and dies or bottoms of the pans, that the pan itself is very little eaten by the acid.

6. *Common Salt* (Chloride of Sodium).—Salt cannot operate directly on the sulphurets. It does not decompose, but it must be first decomposed by another agent, before any action, chiefly chlorination, can take place. Sal ammoniac, chloride of copper, or iron, may replace the salt for the purpose of chloridizing. But, as before mentioned, salt and sulphuric acid create, to some degree, muriatic acid, which acts also on sulphurets.

7. *Chloride of Copper*.—This salt is composed of 52·5 chlorine, 47·4 copper. It is obtained by dissolving metallic copper in *aqua regia*. To this purpose muriatic acid, with some nitric, is introduced into a porcelain or enamelled vessel, and some copper pieces, best in shape of plates or sheets; the acid will dissolve so much of them as to become saturated. Some heat hastens the solution, which assumes a beautiful emerald-green color. The chloride of copper seems to operate with better

result than blue vitriol; but, using this salt with common salt, the amalgam obtained by this process appears often white and clean, crackling between the fingers like pure silver amalgam, still containing a great deal of sulphurets. After retorting, it appears blackish, and renders, by melting, a pure silver bar, but also a rich silver matt, sometimes as much as fifteen per cent. of the bullion, containing eighty per cent. of silver. Such a result is not always obtained, but under some circumstances, not yet sufficiently investigated. When experimented on in small quantities of about twenty-five pounds, and the calculation made also on the silver in the matt, the loss appears between fifteen and twenty per cent. In fact, however, the use of sulphate of copper and salt amounts exactly to the same thing, because the copper vitriol, especially if steam is introduced, is soon decomposed by the salt forming chloride of copper and sulphate of soda.

8. *Subchloride of Copper.*—It consists of 36 chlorine, and 64 copper. It is obtained by boiling metallic copper in copper chloride, as prepared in the above described way. It changes the color from green into brown, but appears light green, when diluted with water, giving a precipitate.

9. *Protochloride of Iron.*—It consists of 66 chlorine, and 34 iron. It can be prepared in different ways. Small pieces of iron are dissolved in muriatic acid to

saturation, supported by a moderate heat. The iron is then removed, or the solution poured over in another porcelain dish, adding so much muriatic acid as was taken for the solution, and heated again, whereupon some nitric acid is added, but carefully, in very small quantities, till the boiling up ceases. It behaves similar to chloride of copper.

10. *Chloride of Iron.*—It is obtained by boiling or heating metallic iron in *aqua regia* or muriatic acid. It acts favorably on the silver sulphurets.

QUANTITY OF CHEMICALS

Per Ton of Ore, as used in different Mills.

Sec. 25. In describing the quantity of chemicals, as used in treating silver ores, it may be observed that, according to the amount of sulphurets, also the quantity of chemicals must be proportionate.

 a. Chloride of copper (Sec. 24. 7)............... 13 pounds.
  Common salt............................. 60 pounds.

 b. Chloride of iron (Sec. 24. 10)................ 13 pounds.

 c. Sulphate of iron........................... 1 pound.
  Sulphate of copper........................ 8 pounds.
  Common salt............................. 80 pounds.

a, *b*, *c*, are calculated for ore containing from two hundred and fifty to five hundred ounces of silver in sulphurets. All chemicals except salt are used in solution.

The salt is charged half an hour before the chemicals are put in.

d.	Sulphuric acid...........................	3 pounds.
	Sulphate of copper.....................	2 pounds.
	Salt.......................................	15 pounds.
e.	Sulphuric acid...........................	2 pounds.
	Alum.....................................	2 pounds.
	Sulphate of copper.....................	1½ pounds.
f.	Sulphate of copper.....................	1·2 pounds.
	Sulphate of iron........................	1·0 pound.
	Sal ammoniac...........................	0·8 pound.
	Common salt............................	2·0 pounds.
g.	Alum.....................................	1½ pounds.
	Sulphate of copper.....................	1½ pounds.
	Salt.......................................	40 pounds.
h.	Muriatic acid...........................	30 ounces.
	Peroxyd of manganese..................	8 ounces.
	Blue vitriol.............................	10 ounces.
	Green vitriol............................	10 ounces.
i.	Common salt............................	15 pounds.
	Nitric acid..............................	1 to 2 pounds.
	Sulphate of iron........................	1 to 2 pounds.
k.	Common salt............................	25 pounds.
	Blue vitriol.............................	2 pounds.
	Catechu.................................	2 pounds.

These, and a great number of other recipes, formerly honored with the title of "Processes," may not all prove to be the results of experience or scientific speculation, especially as to the proportion and quantity. It seems that there is a boundary beyond which the extraction of silver can not be effected, whatever may be done in regard to quality or quantity of chemicals, or in regard to time, if the ore is not roasted. But, as a matter of course, some chemicals work much better than others. Some ores are better suited for the pan amalgamation without roasting than others. The chemicals under *a*, *b*, *c*, and *d*, will likely give the most satisfaction.

AMALGAMATION IN PANS.

Sec. 26. The pans, especially the mullers, have different shapes; but, although the results of operations depend considerably on the right form óf mullers, effecting more or less perfect grinding, it will be sufficient to describe the principal arrangement of the common pans, and of the Wheeler pans, the latter differing entirely from all others except Varney's, and lately Hepburn's.

(For a description of the common pan, see Sec. 49.)

The treatment of ores in iron pans is the most simple amongst all metallurgical operations. The muller is put in motion, the pan charged with some water, and the ore, in pulverized condition, introduced. A four-foot pan may take one hundred and fifty pounds; a five-foot two hundred and fifty or three hundred, and a six-foot pan four hundred and fifty or five hundred pounds of ore. Wheeler's pan of four-foot, is charged with seven hundred and fifty pounds. The quantity of water is soon found out. The ore must be kept in a thickish condition. If there is too much water, there is not only more friction, by the settling of the sand, but also the chemicals are more diluted, and the quicksilver is in one mass on the bottom. But if, on the other hand, the pulp is too thick, the particles of ore cannot change their places quick enough, dead masses will occur, and the amalgamation is delayed.

When the pan is thus charged with ore, the quicksilver is next introduced, in quantity of thirty-five, sixty, or eighty pounds, according to the sizes of the pans above mentioned. If salt is used, it may be added immediately after the quicksilver, and half an hour's time allowed for dissolving, before other chemicals are charged. The sulphuric acid must be diluted with about four parts of water before being introduced. The pulp of ore must be kept as much as possible in a uniform condition in regard to dilution and heat, and no boiling allowed. The right speed of the muller is between ten and fifteen revolutions per minute. A quicker motion is not injurious to amalgamation, unless there is too much water in the pan, or the mullers are not set right, and throw the ore towards the sides. The temperature seems to answer best when below boiling heat, but still hot enough to be inconvenient to hold the finger longer in the pulp than just to try its condition. Too much heat is injurious, especially when sulphate or chloride of copper is used, causing a larger loss in quicksilver, and also in amalgam, which, assuming a black, rag-like appearance, does not unite easily with the other amalgam, but swims on the quicksilver. It parts in minute particles, and is liable to escape in discharging the tailings. Such black amalgam rubbed in a porcelain mortar, gives a great deal of black powder, consisting of some silver sulphurets and sulphide of copper, resulting from sulphate of copper, probably by action on sulphurets of silver.

Treating the ore cold, using sulphuric acid and salt, the result, in regard to the quantity of amalgam, does not differ much; but after retorting and melting, the warm amalgamation gains from five to seven per cent. on the value of the bar. Chloride of copper and chloride of iron allow cold amalgamation, but warm amalgamation may be better for some qualities of ore.

After a run of three or four hours, during which time the ore is ground very fine, water is introduced, and the pulp diluted, so that all quicksilver and amalgam can join in one mass on the bottom, which may require half an hour's time. There are generally three discharge holes to each pan. The lowest one serves for discharging quicksilver and is level with the bottom or even below it, communicating with a groove, which runs to the centerpiece. The uppermost hole is opened and the tailings discharged, under a constant stream of water, into the pan. After one-quarter of an hour the lower plug is removed, continuing the discharge for another quarter of an hour. Both holes are plugged up again, the pan charged with ore and treated as before. The quicksilver may be taken out once or twice a week or oftener according to the richness of the ore.

In discharging the tailings, some amalgam is always

carried out, especially at the end of the operation, when the coarser sand is washed off. To prevent the loss of this amalgam, agitators are applied. They consist of tubs, two or three feet in diameter, and ten or twelve inches high, with a vertical shaft, on which four arms are fixed, having vertical stirrers as represented in Fig. 17. But it is evident that one agitator for many pans, as is the general usage, is not sufficient; it is only a repetition of the first washing. The greatest part of the amalgam goes out again. These agitators differ very much in construction and size, none, however, offer a satisfactory result. The most proper and time-saving way, in discharging the pans where a constant stream of clear water can be obtained, is the use of one or two general agitators, six or eight feet in diameter, in which the tailings are discharged, leaving the quicksilver with a part of the ore always in the pan. The tailings are not diluted, unless the stuff is too thick. In this case some water may be introduced into the pan before discharging, but not so much as to allow a separation of slime and fine sand. From the agitator the tailings run off in a small stream (three-eighths of an inch thick) into a five-foot pan in which a continual stream (an inch in diameter) of clear water flows, as described in Sec. 27. This strong dilution under a constant stream permits a very close separation of amalgam which unites with the quicksilver in the pan. Such pans are of better service, when the continual discharge is arranged in the centre like Knox's pans, because the motion of the muller (fifteen to eighteen revolutions per minute) creates a strong current on the periphery.

When the tailings are discharged, there is always a good deal of sand left in the pan, partly to save time, but chiefly on account of quicksilver and amalgam, of which at the end of the operation, more is carried out in proportion.

Whenever the quicksilver gets thick by the amalgam or at the close of the week, the lowest hole is opened, after the tailings have been removed by the usual discharge. The quicksilver runs into buckets, but some of it remains in the pan and must be taken out with the scoop. By means of a piece of blanket and clear water, the quicksilver is washed perfectly clean, and poured into a filter, of sugar-loaf shape (see Fig. 18), of strong canvas or duck. The quicksilver runs through the cloth, leaving the amalgam in the filter, which generally must be pressed over again through a cloth by hand or a press. This amalgam must be separated, the metal from the quicksilver, by retorting (see Sec. 42).

All the sand, sulphurets, and amalgam particles from the last discharge and cleaning of the quicksilver, must come back into the pan with the next charge. If, however, the sulphurets and metallic iron from the wear of the shoes accumulate too much, forming a heavy stuff, which generally retains a great deal of amalgam, it is better to treat it separately in a pan with addition of one, or one and a half per cent. of sulphuric acid, and to save all these tailings, which contain such silver combinations, as resist decomposition in pans. These tailings, if accumulated to several tons, must be roasted with four or five per cent. of salt, and treated in pans without chemicals.

AMALGAMATION IN WHEELER'S PANS.

SEC. 27. The amalgamation in Wheeler's pans (see Sec. 50, Figs. 24, 25) does not differ much from that of others, but the peculiarity of discharging tailings and quicksilver every time, at once, which is based on the construction and motion of the muller, requires attention to some points, which partly influence the good results of the amalgamation. A quick and fine grinding at the speed of fifty to sixty revolutions depends also on the quality of the shoes and dies. Soft shoes will stand according to the quality of ore thirty to forty days, while hard ones of white iron, will last ten to fifteen days longer, doing at the same time better grinding. The only inconvenience of the hard shoes and dies consists in the brittleness of the iron. On this account they must be removed in time, so that not more than three-fourths of an inch are worn off, leaving one-fourth of an inch in thickness. If this be neglected, and the shoes become thinner, a trifling foreign hard matter may effect the break of shoes or dies, and cause a break in the yoke or something else.

The advantage of a quick and prompt grinding will be found indirectly in saving silver ore, because it allows the use of coarser ore in the pan. It is a known fact that the finer the ore is crushed, especially wet, the more silver ore will be turned into slime, of which a great deal will be carried out by the water (*). Crushing dry, the loss in dust will increase with the fineness of the sieves.

It is therefore advisable to crush coarse and to grind fine. Using common pans, it is preferable to pulverize the ore as fine as possible in order to save time and the shoes of the pan, especially the bottom. Wheeler's pan performs the pulverization much quicker and cheaper than the battery does, at a certain size of grain.

The guide-blades must be kept always close to the muller, in order to prevent ore and quicksilver from following the motion of the muller, thus forcing the mass to the centre and under the muller. The present construction is such, that the muller cannot follow the wear of the shoes for more than about half an inch. The grinding becomes imperfect. In this case, which may happen once in a week, the key and screw which fasten the yoke to the driving shaft, must be loosened, and the shaft raised half an inch, paying attention that the muller, after all is fixed, can be raised above the dies at least one-eighth of an inch. When this is done, the guide-

---

* That the slime is very difficult to work in pans to advantage is known to every millman. On this account and because the slime is often richer than the coarser sediment, especially if brittle silver sulphurets are present, the coarse crushing is preferable. It was even observed that more quicksilver scum is formed in treating slime than in working coarser material.

blades have to be lowered by the screw on top of the shaft, so as to have it again close to the muller. If hard shoes are used, this adjusting may not be required for a longer time. It takes, however, only a few minutes.

The consistence of the pulp must be thickish, but still so, that a lively motion on the sides and between the guide-blades is perceptible. If the motion be too slack, some water must be added. In regard to amalgamation, a little more or less water is not so important as it is for the separation of amalgam in the agitators.

Steam must be introduced three or four inches above the muller, for if too close to the bottom, the pipes require frequent cleaning. The temperature can be kept near to the boiling point. There is no necessity of a steam chamber. The direct use of steam in the pulp does not interfere with the required consistency, but allows a considerable saving of fuel. Eight Wheeler's pans, if the boiler is proportionate, require half a cord of wood in twenty-four hours.

In regard to chemicals, after several months' run at Col. Raymond's mill on Carson River, experience proved that the use of chemicals is entirely useless in treating Ophir ore. I made comparative runs for many weeks, and found that experience confirmed that a better result was obtained without any chemicals whatever. The use of them was therefore abandoned at the Dayton mills. It is, however, a fact, that in treating silver ore in common pans, the use of chemicals always gives a better result. It can not be supposed, therefore, that the chemicals prevent a perfect amalgamation in Wheeler's pans, but the amalgam is in a different condition, more liable to be ground into a black floating powder. This is also the reason that it seems as if the quicksilver were protected by the chemicals against being ground to scum, because the black powdered quicksilver is less visible.

The loss of quicksilver could not have been ascertained yet, as it is not advisable to take out the dies on account of cleaning. As a matter of course, the quicksilver must be turned into scum to some degree, but considering the better yield and expelling of chemicals, these pans can afford to lose some quicksilver. In the course of manipulation, however, a great deal of this scum is regained.

It is unquestionable that the iron under favorable circumstances decomposes the silver sulphurets just as well or better than the best chemicals; but if, on account of improper arrangement of grinding, the action of iron on sulphurets is effected in a limited degree, the chemicals will then assist. It happens often that in cleaning the agitator, when no chemicals are applied in the pans, a strong smell of sulphuretted hydrogen is observed, which arises from the decomposition of sulphurets. In washing the tailings, after the iron has been extracted with the magnet, blue sulphurets can hardly be discovered. The difficulty in stating the loss of silver lies partly in the inconvenience of taking out the dies, in order to effect a perfect cleaning, as it is almost impossible to place the dies back again as level as before, but the difficulty is chiefly found in the attempt to get a reliable average sample for the assay, especially if wet crushing is going on. The sulphurets and gold are always concentrated in the vats, where the ore falls in from the battery, and every inch in it shows a different amount of silver. The next vat shows the same difference, having, besides, already so much slime that a mixing for the purpose of having the metal equally distributed in the mass, is impossible without drying.

The calculations on the superiority of Wheeler's pans, at least to the present date, are derived from the comparative yield and the appearance of the tailings, according to which these pans seem to yield at least ten per cent. more than the best common pans. To find the real loss of silver and quicksilver, it would require dry ore, each charge well mixed, sampled, and weighed, and one pan and agitator exclusively for this purpose. This could not have been performed yet.

A great difference as to the result will be found in a comparative working of the finest sediment (the slime) which glides between the mullers of the common pans almost motionless, as if it were one mass, while in Wheeler's pans the slime is forced under the muller with the same speed as the ore.

When the muller is in motion, so much water is introduced that it plays over the rim of the muller; then seven hundred and fifty pounds of wet ore, or six hundred of dry, are charged, and if it is perceived that the motion of the ore is not lively enough, more water is added. The pulp should not cover the guide-blades. After this, one hundred pounds of quicksilver are poured into each pan. The amalgamation and grinding continue for three hours. A longer amalgamation does not appear to give a better result. The plug of the hole which is nearest to the bottom is taken out, and the whole mass discharged. The pan is charged immediately as before. The charge and discharge can be effected in five minutes. Of the hundred pounds of quicksilver, which were introduced, only fifty or sixty pounds were discharged with the ore; the balance remains between the dies. All the subsequent charges take fifty pounds. Several hundred pounds of quicksilver must be kept at hand, to replace such as remains with the amalgam in the agitator.

Wheeler's agitator (see Sec. 51), being eight feet in diameter, has such a swift motion on the periphery, that, treating the tailings in the usual way, by diluting with clear water, a great deal of amalgam is carried out. The discharge from the upper hole, where the mud and very fine sand with much water flows out, is free of amalgam, but after all the mud has been removed and the lower hole opened, the floating amalgam, which cannot reach the bottom on account of the sand (which settled in consequence of the dilution), escapes with the tailings. The loss increases with the speed and also with the coarseness of the tailings. If the speed is reduced, the motion near the centre will be too slow.

After different experiments I adopted the following method, by which the above difficulties are avoided: Three inches above the bottom (see Fig. 23) there is a three-eighth-inch pipe, through which the tailings flow out in a thin stream continually, so that in the course of three hours, if three pans are discharged into one agitator, the tailings are down to the level of the pipe. The quicksilver and amalgam have time enough to sink by degrees to the bottom, and those particles which escape through the three-eighth-inch pipe are also saved.

No water is needed in the agitator, but the pulp must have the proper consistency when discharged. The amalgam accumulates round the centre-bowl. It will be found that this amalgam, and that from the bowl, is richer in gold than that further off, while the fine amalgam carried out through the pipe contains only one-half of the amount of gold which is found in the average of the agitator amalgam. It occurs often that a great part of the amalgam deposits in the amalgamating pans, sometimes accumulating on the sides, and accidentally on the muller, which, if perceived at the discharge, must be taken out, else it might accumulate to fifty pounds, or more, pressing the muller on one side, and thus causing a very unequal wear of the shoes. At other times, no deposit of amalgam takes place in the pans, or one pan may retain as much as one hundred pounds, when another has none at all. It seems that this appearance depends on a variable electric condition of the iron.

When the pans are discharged, the quicksilver runs on the inclined bottom to the centre, joining the quicksilver in the bowl, coming out by the siphon, whence it is taken and returned to the pans, fifty pounds to each charge. The amalgam accumulates also in the bowl, preventing finally the passage through the siphon. In this case the agitator must be cleaned. To this purpose, when the last discharge is effected, the lowest one-half inch pipe is opened, and the tailings, when level with the half-inch pipe, are discharged by the hole on the inclined bottom near the bowl. The tailings from this last discharge are led into a separate box, whence it can be transferred immediately into the other agitator, or if only one, back into Wheeler's pan. The agitator is then stopped, the black stuff from the bottom, which is very rich in fine amalgam, removed, the quicksilver strained, and the bowl filled again with quicksilver after the siphon was also cleaned. All this can be done in three hours, so that the agitator is ready for the subsequent discharge of the pans. The shoes of the agitator must be run suspended about one-eighth of an inch above the bottom, in order to allow the settling of sulphurets and amalgam.

If the ore contains a great deal of sulphurets of such a nature that a part of them do not yield up their silver without roasting, the agitator may be used as a concentrator at the same time, by screwing up the shaft, on the arms of which the shoes are fastened, for one-eighth of an inch or more every day.

All the black stuff from the agitator must be worked over in a common five or six-foot pan without the addition of quicksilver, and the tailings saved to be roasted.

Each agitator must have one four or five-foot pan for the reception of the tailings, coming out of the one-eighth inch pipe. The discharge-hole of this pan, about three inches above the bottom, is always open, so that the tailings have only a passage through the pan, the muller of which may make twelve or fifteen revolutions. The fine amalgam, and especially the quicksilver scum, would escape from this pan to the greatest part, if not diluted. A continual stream of clear water, therefore, through a pipe an inch or three-quarters of an inch in diameter, is indispensable. Twenty or thirty pounds of quicksilver are introduced into the pan in order to collect the amalgam.

In some mills Wheeler's pans are discharged, using only the upper hole, retaining thus the most of the quicksilver. This mode saves the trouble of handling the fifty pounds required for each charge; but the amalgam, accumulating in the quicksilver, is too much exposed to grinding, causing thus a richer quicksilver scum, of which a considerable part is lost.

Wheeler's pans have been improved lately by Mr. Hepburn. The improvement seems to be important. Without having a larger diameter, the conical bottom offers a larger grinding surface. The whole arrangement is simplified. On account of the inclined bottom, no guide-blades are required. One of these pans is charged with 1,000 pounds, so that four tons of ore may be worked in twenty-four hours by one pan, requiring about two and one-half horse-power each. In regard to the amalgamation, there will be probably the same result as obtained by Wheeler's. All that was said in relation to the manipulation with Wheeler's pans can be applied also to Hepburn's.

II. ROASTING PROCESS.

Sec. 28. The roasting of silver ore, for the purpose of converting all the silver into a chloride, no matter in what condition it may occur, either in the metallic state, or in a sulphuret, arsenide or antimonial silver, in order to make it fit for easy decomposition and subsequent immediate amalgamation, has been adopted in several extensive works of Nevada Territory. There may be some kinds of silver ore, which under certain circumstances can be treated raw in pans very advantageously, even if there be a greater loss of silver. Nevertheless, the high importance of roasting can not be overlooked, this being in many instances the surest and the only way to beneficiate such silver combinations as refuse to deliver the silver by amalgamation without roasting. There is no silver ore which can not be treated successfully by means of roasting. The patio or pan amalgamation, working so-called "rebellious" ores, must resort to roasting, or else exclude such ores from the manipulation.

In regard to the importance of roasting, which in course of time may be more appreciated, a thorough description of this process appears necessary. Moreover, it is entirely impossible to carry on a rational and correct roasting according to the quality of ore, or in regard to a special object, if the chemical actions during the roasting are not regarded or understood. It is indispensable to get acquainted with the theory of this process, of which the most important actions will be considered.

Three agents are active in roasting: the oxygen of the air, the hydrogen of the water-vapors of the air and fuel, and the chlorine of the salt. (By the term "roasting" I here always mean "chlorination roasting.") But as the decomposition of salt by heat alone is very imperfect, sulphur is another important agent, the presence of which is by all means required. The decomposition of salt is effected by sulphuric acid in the form of vapor, and by sulphates, produced by the oxygen of the air and the sulphur of the sulphurets in the ore. If, therefore, the silver were combined with antimony or arsenic, with but little or no sulphurets, the chlorination would be very imperfect. In this case an addition of two or three per cent. of calcined green vitriol (sulphate of iron) is required.

The principal object in roasting must be first, the production of sulphates. The iron pyrites and other sulphurets, when red hot and acted upon by the oxygen of the air, change into sulphates by oxydation of the sulphur to sulphuric acid. The sulphuric acid can not unite with the metal of the sulphuret under decomposition, unless it becomes an oxyd. One part of the sulphuric acid, therefore, transfers oxygen to the metal, being thus reduced to sulphurous acid, while the metal becomes an oxyd and combines with another part of sulphuric acid to a sulphate.

During this process, which is performed at a low heat, the salt is almost entirely indifferent, so that it is immaterial whether it is charged at the same time with the ore, or two hours later.

As soon as the sulphates are formed, and no odor of sulphurous acid observed, the heat must be increased. The decomposition of salt begins, being performed in two different ways:

1. The sulphate of iron principally, and other sulphates, emit sulphuric acid in vapor, which, in contact with salt, forms sulphate of soda, setting free the chlorine in a gaseous form. One part of the oxygen of the sulphuric acid is transferred to the sodium of the salt, oxydizing it to soda, which, uniting with another part of sulphuric acid, forms sulphate of soda, while the chlorine absconds, combining with free metals in chlorides when in contact with them, and decomposes sulphurets in such a way, that one part of the chlorine combines with the sulphur in volatile chloride of sulphur, while

another part unites with the freed metal of the sulphuret in a chloride.

2. The other way of decomposition of salt differs in its result, not emitting chlorine gas, but forming chlorides during the act of decomposition. The sulphate in contact with salt enters into an exchange of compounds. The sulphuric acid combines with the soda of the salt to sulphate of soda, and the chlorine with the metal of the sulphate to a chloride, the oxygen of the metaloxyd oxydizing the sodium. The chlorination of metals accordingly is performed by the direct action of chlorine gas on the metals and sulphurets, and by contact of salt with sulphates.

During this process, besides the chlorine gas, hydrochloric acid in vapor is created. The hydrochloric or muriatic acid arises partly by the action of water-absorbing sulphuric acid on the salt, whereby the sodium is oxydized by the oxygen of the water, while its hydrogen unites with the chlorine to form hydrochloric acid in the form of gas. The muriatic acid is also formed by the contact of chlorine gas with compounds of hydrogen, for instance carburetted hydrogen. The chlorine, by its affinity for hydrogen, decomposes the compounds of the latter elements. It is also produced by the contact of steam with volatile chlorides, as chlorides of antimony, zinc, lead or copper, etc., reducing the metals to oxyds, or the silver to a metallic state, which, however, in contact with hydrochloric acid or chlorine, is turned again into a chloride.

The circumstances, accordingly, under which the formation of hydrochloric or muriatic acid takes place, are various, but always when water-vapors enter the porous mass of ore.

Behavior of Chlorine Gas.

Sec. 29. It has been mentioned already that chlorine gas acts directly on sulphurets. Under the action of chlorine gas the following changes occur:

a. The *Iron* (with sulphur or arsenic) changes into protochloride of iron ($Fe\ Cl$), but, exposed to the air, into sesquichloride ($Fe^2\ Cl^3$). This chloride becomes volatile and is sublimable. If in this condition it meets gaseous products of burning fuel containing vapors of water, or hot air containing steam, a mutual decomposition takes place, resulting in oxyd of iron and gaseous hydrochloric acid.

b. *Manganese* (combined with sulphur) changes into protochloride of manganese ($Mn\ Cl$). It is not volatile. Water-vapors decompose it into sesquioxyd and gaseous muriatic acid.

c. *Zinc* (combined with sulphur) changes into protochloride of zinc ($Zn\ Cl$). It melts before it is red hot, and becomes volatile when red hot. In contact with steam, it forms oxyd of zinc and hydrochloric acid.

d. *Lead* (in combination with sulphur) changes very

slowly into chloride of lead (Pb Cl). It melts easily. In contact with red hot air it evaporates partly, while another part, evolving chlorine, changes into a compound of oxyd of lead and chlorid of lead, which is not volatile.

e. Copper (combined with sulphur) changes partly into sesquichloride of copper (Cu^2 Cl), partly into protochloride (Cu Cl), according to the action of more or less chlorine at a higher or lower temperature. Both combinations are inclined to evaporate. When red hot, the chloride changes into sesquichloride, emitting half of its chlorine, by which sulphurets are decomposed. Under the action of steam a mutual decomposition takes place, creating gaseous hydrochloric acid and oxyd, or sequioxyd of copper, the latter being converted into oxyd by contact with the air.

f. Silver (native and in combination with sulphur) changes slowly into chloride of silver (Ag Cl.) It becomes volatile only at a high temperature.

g. Gold (free or combined with arsenic, antimony, or tellurium) changes when in a very fine pulverized state, at a low heat, into terchloride of gold (Au Cl^3). It emits two parts of chlorine, below red heat, forming chloride of gold (Au Cl). Red heat changes it into metallic gold.

h. Arsenic (with other metals and sulphur) is transformed into a very volatile terchloride of arsenic (As Cl^3).

i. Antimony (with other metals or sulphur) changes into terchloride of antimony (Sb Cl^3). It is like the terchloride of arsenic, very volatile.

Behavior of Hydrochloric Acid.

Sec. 30. The gaseous hydrochloric acid in contact with metallic silver unites with it at a high temperature to form chloride of silver. The hydrogen is set free. It behaves in like manner with the sulphurets and arsenides, of which the most are decomposed in such a way that chlorides of metals are formed, while the sulphur or arsenic combines with the hydrogen.

Behavior of Salt.

Sec. 31. When the roasting of ore has advanced so far that considerable quantities of chlorides, which are partly volatile, are formed, under the action of chlorine, and hydrochloric acid in contact with salt and sulphates, some of the salt, not previously decomposed, evaporates. These salt vapors, and those of the volatile chlorides transfer chlorine to undecomposed sulphurets or arsenides, or to already present sulphates, arsenates, or antimonates, or to free oxyds. Chlorides, which are disposed to transfer chlorine to such metals in combination with sulphur or arsenic as possess more affinity to chlorine, than themselves, are, besides salt, protochloride of iron, protochloride of copper, also the chlorides of zinc, lead, and cobalt.

a. Metallic Silver in contact with salt changes partly into chloride of silver, probably in such a way that the silver decomposes the salt in the same proportion as the sodium takes up carbonic acid from the gaseous products of burning fuel.

b. Sulphurets in contact with salt are not decomposed directly. The sulphurous acid, however, in contact with the air, creates sulphuric acid, which acts on the sodium, freeing thus the chlorine, by which the formation of chlorides in the not yet decomposed sulphurets are effected.

c. Arsenides are not changed by the salt. They oxydize, evolve arsenous acid, and are converted into arsenates. Only a very slight decomposition of salt takes place. The presence of sulphates, however, volatile chlorine, or gaseous hydrochloric acid, effects the chlorination.

d. Oxyds of Metals, with the exception of the oxyd of silver, are changed very little or not at all by salt. The oxyd of silver readily giving up its oxygen, changes perfectly into chloride of silver, if sufficient salt be present. A small portion of the oxyds of copper and lead are changed into chlorides.

e. Sulphates.—Sulphates decompose the salt by mutual exchange of compounds. The sulphate of lead changes into chloride of lead, which, evaporating in contact with air, emits one part of its chlorine, being reduced to a combination of chloride and oxyd of lead. The sulphate of copper changes into chloride of copper. This becoming volatile, evolves chlorine gas, and forms sesquichloride of copper, which is less volatile.

Sec. 32. The ores intended for roasting must be examined not only in regard to the quality and quantity of sulphurets, but also in regard to the earthy matters accompanying the ore. It is not immaterial whether the ore contains carbonate of lime or quartz. If there is a great deal of lime in the ore, it absorbs sulphuric acid, forming sulphate of lime, remaining in this condition through the whole process, without being decomposed. On this account calcareous ore requires so much more sulphurets or sulphate of iron as is necessary to change all the lime into sulphate. Talcose ores behave like the calcareous. Silicia or quartz, if abundant, in presence of steam decomposes some of the salt, when red hot, forming silicate of soda and hydrochloric acid, the importance of which has been mentioned (see Sec. 30). This behavior of these earths shows that it is disadvantageous to submit pure calcareous or talcose ores to roasting, and that in such a case quartzose ore must be added, if possible.

The quantity of sulphurets in the ore is important, a

certain amount of it being required to decompose so much salt as is necessary for chlorination. In Freiberg (Germany), it was the rule to subject only that ore to roasting which contained enough sulphurets to give twenty-five or thirty per cent. of matt (sulphide of iron), when assayed for that purpose. If less matt was obtained, the ore had to be mixed with other ore, or so much iron pyrites was added that the required quantity of sulphurets was obtained. The second class ore of the Ophir and Mexican claims in the Comstock lode, consisting of pure decomposed quartz, contains silver sulphurets, with a small proportion of iron pyrites, yielding from six to eight per cent. of matt. The roasting with salt, however, gives a satisfactory result, which must be attributed chiefly to the pure quartzose condition of the ore.

If the ore contains an abundance of sulphurets, the roasting must be performed without salt, for about two hours, till the greatest part of the sulphur is driven off, otherwise it would bake, and cause an imperfect roasting.

The quantity of sulphurets has a great influence on the result of roasting. Ore like that of the Ophir or Mexican mines, containing silverglance, polybasite, brittle silver ore, native silver and gold, some iron, and but little copper pyrites, will give a good result by roasting, even when less attention is paid to the time and diligent stirring, than for instance with the so-called " base metal ore," which abounds in copper pyrites, zinc-blend, sulphuret of lead, etc. The presence of base metals causes a higher loss in silver. The chloride of silver is not volatile, except at a high temperature (Sec. 29, f). But it has been observed that, in the presence of base metal chlorides, the chloride of silver volatilizes also. The increased heat increases the volatilization, but decomposes the base metal chlorides. By keeping a low heat, the loss of silver is less if the zinc-blend is not argentiferous, the latter requiring a higher heat to effect decomposition. But in roasting at a low heat, the base metal chlorides remain in the ore, and cause more loss of quicksilver in the subsequent amalgamation, and require more metallic iron in the barrels; besides, the bullion contains a great deal of base metals. In treating such ore in the roasting furnace, the application of steam is advantageous, creating hydrochloric acid by the decomposition of chlorides, at the same time becoming a decomposing agent for the sulphurets. The hydrogen of the steam decomposes also the chloride of silver, which, upon being reduced to a metallic condition, by its affinity for chlorine, in turn decomposes the hydrochloric acid. The silver may thus change repeatedly from metallic condition to the chloride, while the base metal chlorides are reduced to oxyds, and in that state do not interfere with the amalgamation.

A. ROASTING OF SILVER ORES

For the Barrel and Veatch's Steam Amalgamator.

SEC. 33. The silver ore for the modes of amalgamation without friction, as shown by long experience, must be free from metallic gold, or it must be extracted before the ore is subjected to roasting, as is done in the "Silver State Reduction Works," on Carson River. After roasting, the gold is not like silver, in a soluble and easily decomposable condition, but in a metallic state, generally coated with some oxyd, especially if sulphuret of lead occurs in the ore. This renders the amalgamation of gold much more difficult. Also, metallic silver, when roasting is performed without salt, using charcoal and saw-dust for the purpose of converting the silver-combinations of the ore into the metallic state, will be imperfectly amalgamated in the barrels.

The ore does not require to be very fine for the purpose of roasting, but it must be fine on account of amalgamation. Vertical wirecloth sieves at the battery, with nine hundred holes to the square inch, if dry crushing is in use, or sixteen hundred holes when wet crushing is preferred, on account of extracting gold, will answer the purpose.

The pulverized dry ore is spread on a platform and mixed with from six to twelve per cent. of salt, according to the richness and quality of the ore. It seems, however, that six per cent., as used in one of the

Washoe valley works for ore, assaying from seventy-five to one hundred dollars per ton, may not be sufficient, still the result is considered quite satisfactory. Very rich ore may require fifteen or twenty per cent. of salt. In dry crushing, the salt may be mixed with the ore in the right proportion before going to the battery. This mode of mixing is the most perfect and most convenient.

The Red Hot Furnace (see Sec. 53, Figs. 29, 30) is charged with eight hundred or one thousand pounds of ore, and, by means of iron hoes spread over the bottom of the furnace. If the ore is moist, or rich in sulphurets, the heat is kept low. The workman commences to stir the ore with the hoe or an iron rake, back and forward across the hearth, moving slowly from the bridge towards the flue and back. When the ore is perfectly dry, appearing very movable, almost flowing, the heat must be increased. Continual stirring is required, in order to expose new ore on the surface, thus facilitating the oxydation. If there is a great amount of sulphurets in the ore, the sulphur commences to burn when the ore gets dark red hot, evolving so much heat that the firing must be suspended for about one hour and a half, but the stirring continues, touching all spots and corners of the hearth. After a great part of the sulphur is burnt off, the temperature will sink, and the ore appear dark. The temperature must be raised again by firing. The formation of sulphates is going on, disengaging a large quantity of sulphurous gas. The ore at the bridge will be heated much more than on the opposite side. The roaster must take the trouble of changing the ore from the bridge to the flue, and the cooler ore to the bridge, several times. If lumps are perceived in the ore, they must be beaten to powder by an iron hammer-like instrument with a long handle. After three or four hours roasting, according to the amount of sulphurets, no sulphurous acid is perceptible. The temperature must be increased to a light red heat. The formation of sulphates, arsenates, antimonates, and oxyds is almost completed. The chlorination has commenced, and as the increased heat is rapidly going on, white fumes arise, and the gases and vapors evolved have a sharp, acrid odor, consisting of some sulphurous acid, chlorine gas, hydrochloric gas, chloride of sulphur, chlorides of iron and copper, etc.

The ore assumes a spongy or woolly condition, increasing in volume. In the presence of sufficient copper the flame is colored blue by the chloride of copper. After one hour's roasting, at an increased heat and with diligent stirring, the chlorination is finished. The ore is discharged by the back door or the discharge hole in the bottom, although the fumes and gases are still being evolved.

If there is a great deal of copper and other base metals in the ore, the roasting may require more time in order to decompose the chloride of copper and sulphates, the presence of which in the amalgamating barrels or tubs destroys not only more iron, but increases the heat too much, causing an injurious division of the quicksilver into small particles and scum. The base metal chlorides, reduced by the iron, enter into the amalgam and make it impure. The time for decomposition must be prolonged in the barrels before the quicksilver is introduced, otherwise a destruction of mercury would follow.

The decomposition of the base metal chlorides can be effected in the furnace either by carbonate of lime or by heat, the latter requiring more time. The carbonate of lime in pulverized condition decomposes the chlorides and sulphates, but not the chloride of silver. The addition of lime rock, after the heat has been increased, must be made gradually in regard to the quantity, commencing with two per cent., till the required amount for a certain class of ore is found. It may require as much as six per cent. The first portion of lime is introduced by a scoop, spreading it over the ore and well mixed.

A small portion of the ore is then taken in a porcelain cup or glass, and mixed with some water by means of a piece of iron with a clean metallic surface. If the iron is coated red with copper, or if the water is bluish, some more lime is required. After the lime is charged, half an hour must be allowed for reaction. When another test does not show the above signs of soluble copper, or only in a slight degree, the charge can be taken out. In the absence of lime, wood ashes may be used. If too much lime or ashes is used, the amalgamation is injured, and a greater loss of silver will be the result. The chloride of copper will also be decomposed by longer roasting and increased heat, and samples should be taken in the same way as before to ascertain the decomposition of those chlorides.

One furnace requires two men by day and two men by night in order to keep up continual stirring, firing, charging, and so forth. The Central mill at Virginia City employs two men at a time, attending three furnaces. The stirring is performed at intervals; roasting six or seven hours, in treating rich ore. Other works having poorer ore finish the roasting in four and a half and five hours.

The ore after having been roasted contains from five to fifteen per cent. of lumps, which are not roasted thoroughly, and contain some undecomposed sulphurets, sulphates, and chlorides. These lumps are separated from the fine, well-roasted ore, pulverized, and with the addition of two or three per cent. of salt are roasted again for two hours. The ore is sifted through two sieves. One has 64, the other 2,500 to 3,600 holes to the square inch. The lumps which do not pass the coarse sieve are pulverized under stamps and reroasted. The ore which passes the first sieve and stops at the second is ground fine and delivered with the fine sifted ore for amalgamation.

In the Central Mill there is no grinding after roasting. The ore is pulverized under a set of small stamps and sifted while the coarser particles are constantly elevated to the battery.

B. ROASTING OF SILVER ORES

For Pan Amalgamation.

SEC. 34. The roasting of ore for the purpose of amalgamating in iron pans, differs from the already described procedure, in so far as a perfect chlorination of all the silver in the ore is not absolutely required. Consolidated fragments or lumps formed during the roasting are not injurious. The extraction of gold before roasting is not necessary.

The sulphates, remaining in the lumps, and such as were not changed into chlorides by improper roasting, are partly decomposed in the pan by the iron, but most of them are converted into chlorides by the salt, which always remains in small quantities in the ore after roasting. This salt dissolves in the pan and changes the sulphates, which are also soluble, into chlorides, they being decomposed by the iron and amalgamated. The quicksilver, when present in the pan, takes also part in the decomposition, being thus converted into subchloride of mercury or calomel, which, unlike chloride of mercury, not being decomposed by the iron, causes a loss in quicksilver.

The sulphate of silver, soluble in hot water, will be decomposed by iron into metallic condition, combining with the quicksilver to an amalgam. It would appear, therefore, that roasting without salt, for the purpose of producing sulphate of silver, which is easily beneficiated in the pans, would be more economical by saving salt. It is, however, very difficult to transform all the silver into a sulphate. The sulphates of iron and copper must be formed before the sulphuret of silver can be changed into a sulphate, and if there is not sufficient sulphuric acid emitted by other sulphates, a great deal of the sulphuret will be decomposed into sulphurous acid and metallic silver, the presence of which must be avoided. If arsenic and antimony be present, arsenate and antimonate of silver, which will escape the amalgamation, will be formed. A great part of the arsenate and antimonate of silver will be changed into sulphate of silver, but not all, especially if the ore is poor in iron sulphurets. On the other hand, if the heat is kept too high, the sulphate of silver will be reduced to a metallic state, which, as before remarked, must be avoided; because, while the sulphate is not volatile, the metallic silver, by means of oxydation, evaporates, and deposits itself in cooler places in a metallic condition, emitting oxygen, causing thus a loss.

The chlorodizing roasting requires less attention, and gives a better result. The sulphurets, which may remain in the ore undecomposed after roasting, will be reduced in the pan by predominant chlorides and sulphates. Sulphates alone effect very imperfect decomposition of sulphurets in the pan.

Mr. Sutro's furnace for this purpose (at Dayton, N. T.) is twelve feet by thirteen, offering about one hundred and fifty square feet of hearth surface. The furnace is charged with 2,000 pounds of ore with two per cent. of salt, or when pan-tailings are subjected to roasting, the ore of which had been treated with chemicals and salt, the latter is not added at all. Two men are employed at a time at each furnace for twelve hours. The stirring is kept up constantly at a low, dark red heat four hours long, when the ore is considered well roasted and withdrawn. It contains a great many lumps, so that sifting and pulverizing are required, chiefly on account of the imperfect pan arrangement. In using Wheeler's or Hepburn's pans, it is not necessary to pulverize the roasted stuff. The lumps are not formed in the furnace, but are the consequence of the fine muddy condition of pan-tailings. In roasting dry ore, or well dried and pulverized pan-tailings, the lumps are formed in small proportion. This roasted ore is then introduced into the pan like unroasted ore, and amalgamated in the usual way. A great deal of the base metals will enter the amalgam, if such occur in the ore, but a strange appearance is the iron amalgam which is always obtained in a certain quantity in treating pan-tailings. It separates in melting, swimming on the fused metal in lumps, when it must be removed and melted over with more fluxes. But it sometimes happens that all the amalgam after retorting appears black, spongy, and very light, containing from forty to fifty per cent. of iron. In this case the result in regard to silver extraction is unfavorable. This black retorted iron amalgam, only the result of certain old pan-tailings, must be worked over, treating it like ore, in a pan with quicksilver and some sulphuric acid. The iron entering the amalgam is derived principally from the wear of dies and shoes, being in metallic condition, but after six hours' roasting, most of it is finally converted into an oxyd. Some chloride of iron or other combinations may still remain in a very limited proportion. It is therefore difficult to account for the reason why, and under what conditions the iron is amalgamated, as experiments on ores containing metallic iron from the stamps, when treated in the pans with the addition of sulphate, protochloride, or chloride of iron, always produces amalgam free of iron. At the moment of amalgamation, the iron is in metallic condition, and also after retorting, but when the retort is opened and the air comes in contact with it, the amalgam assumes a higher glow, and continues so for twenty-four hours. During this time, most of the iron will oxydize, and still be attracted by the magnet.

It is, however, very likely that by proper roasting, of a reasonable charge, this singular appearance can be avoided.

As a matter of course, only those ores or tailings which contain a sufficient quantity of sulphurets, especially iron sulphurets, can be subjected to roasting. If the quantity be insufficient, one or two per cent. of calcined green vitriol (sulphate of iron) must be added, or the ore must be concentrated, saving the tailings in the usual way. The concentrated ore, after roasting, may be amalgamated with the unroasted tailings, for which the roasted part represents the chemical.

It is advisable to use not less than four per cent. of salt, and not to charge more than 1,000 pounds at a time, except in a mechanical furnace. The temperature must be kept at a dark red heat for at least two hours, and one hour light red hot, in which time the roasting of 1,000 pounds of ore may be generally completed. In using dry ore, the formation of lumps is moderate, and requires no sifting or grinding, especially if Wheeler's pans are used. I have made different experiments with roasted ore in the pans, always obtaining the best results and clean amalgam, except in one instance, purposely applying a very low temperature, on which occasion, the amalgam, apparently pure, turned black after retorting, and consisted mostly of iron.

But when the ore contains a great deal of copper and other base metals, the roasting must be treated more carefully (Sec. 33). The result of this manipulation is a metal more or less impure, between 600 and 700 fine— that is, if the roasting has been properly conducted and the ore not overloaded with base metals.

C. ROASTING OF SILVER ORES

ABOUNDING IN ANTIMONY, FOR PAN AMALGAMATION.

SEC. 35. Ores, containing an abundance of antimony, like that of the Sheba lode (Sec. 16, 11 b'), which is rich in silver, can not be treated in pans without roasting. This ore is accompanied by sulphuret of zinc, sulphuret of lead, and carbonate of lead. The carbonate of lead is black, somewhat dull, and also rich in silver.

In selecting the ore for this purpose, the sulphuret of lead must be separated as much as possible. Some of the gangue should be left in the sulphurets. Quartz or other earthy matter prevents the baking of the ore while roasting.

The hot furnace is charged with six or seven hundred pounds of ore, and while the mass is kept at a very low temperature, below glowing heat, it is diligently stirred. The sulphurets of antimony and lead fuse at a dark red heat, and if they were fused, the roasting would be most imperfect, and result in a loss of silver. Care must be taken to keep the temperature low, especially when the ore is rich in sulphuret of antimony. The oxydation of the sulphur and antimony will soon commence. White fumes of antimonous acid arise, gradually increasing, a strong odor of sulphurous acid is emitted, while constant stirring exposes always a new surface of the ore to the oxydizing air.

As soon as it is perceived that the fumes and the formation of sulphurous acid decrease, the heat must be raised gradually, so that about two hours after the charge the ore appears red hot. Sulphates of lead and zinc and some sulphate of silver will be formed with the increasing heat, also antimonate of silver, of which only a small part may be changed into sulphate of silver under the influence of the limited quantity of gaseous sulphuric acid. This acid is partly disengaged from the sulphates of lead and zinc under the increasing heat, which at the expiration of three or three and a half hours must be nearly light red. At this time samples must be taken from the furnace, and if ascertained by odor that none or not much sulphurous acid is emitted, the first part of the roasting is finished.

During this period a great deal of antimony is volatilized, as antimonous acid, and also some oxyds of lead and zinc.

One part of the antimonous acid combines with antimonate, which is not volatile. To disengage this, as well as the antimonate of silver, the roasting must be changed into a chloridizing one. For this purpose, five per cent. of salt, in a fine pulverized condition, is thrown into the furnace with a scoop, in such a way as to scatter it over the whole surface of the ore.

Soon after the ore and salt are mixed by the usual stirring, white fumes will arise again, consisting chiefly of chloride of antimony which is very volatile. Also some chloride of lead and zinc are volatilized. The sulphate of silver is changed into a chloride, partly by the decomposition of salt, partly by the volatile chlorine. The formation of hydrochloric acid is here important to assist the decomposition of the antimonate of silver, for which purpose the introduction of some steam in the furnace, under a pressure of three or four pounds, will render good service. The ore increases in volume a great deal, becoming woolly, changing its color by degrees to light yellow. After the addition of the salt, the temperature must be increased a little to a light red heat, and after one hour's chloridizing roasting, the process is completed and the ore discharged.

As a matter of course, some silver is lost, as the chlorides of lead, and zinc, and antimony will dispose the chloride of silver to evaporate. The amalgamation of this ore will yield a metal, containing a considerable amount of lead, according to the quantity of lead in the roasted ore.

Mr. Sutro has used a very simple way of separating the lead amalgam from the silver amalgam. It is known that the silver amalgam, which is obtained in the pan amalgamation, consists of small regular crystals which are suspended in the quicksilver. The lead amalgam, on the contrary, is entirely dissolved in the quicksilver when hot. If, therefore, the quicksilver is pressed through a cloth while hot, the dissolved lead amalgam is found in the quicksilver, and the silver amalgam in the cloth. A second filtration of the quicksilver when cold, gives the lead amalgam. The lead contains three or four per cent. of silver. It is, however, not likely that this way of separation would answer when Wheeler's pans are in use, for they yield a finer amalgam; besides, it is dangerous to handle hot quicksilver. A surer and more perfect separation of lead and silver is effected by refining. (Sec. 47.)

D. ROASTING OF SILVER ORES

In a Mechanical Furnace.

Sec. 36. The difference between roasting in a common and a mechanical furnace is merely a difference of mechanical operation, but on the more or less proper execution of the mechanical operation depends also the chemical result. Gurlt, in his remarks on the new progress of the copper process in England, speaks of a mechanical double roasting furnace with revolving stirrers, which he saw at the Pembrey Copperworks. He says, one of these double furnaces roasts twenty-four tons of ore in twenty-four hours, and recommends it highly for roasting copper and lead ores, but he thinks it would not answer for roasting where a great deal of attention is required, for instance in roasting silver ore. This remark may be true in Europe, but we, in Nevada Territory, are differently situated. In Germany the roaster works with hands and head, and is responsible for the result. His work can be trusted. Our roasters are inexperienced, frequently green hands, without the least interest in the result. A good mechanical furnace is also reliable in its performance, and responsible for the result.

The mechanical furnace which I propose is not in use to my knowledge, but it may be easily perceived that no other furnace offers the same advantage of having the ore so uniformly heated. Whatever the construction of the roof or arch of a furnace may be, the heat will always be more intense at the bridge than on the opposite side. This requires the troublesome moving of ore from the bridge to the flue and back, and even then the disadvantage of the difference in temperature is not entirely corrected. The revolving bottom of the furnace (Sec. 54, Fig. 31) carries the ore at each revolution through all the different temperatures of the furnace, and the ore is twice stirred, thus effecting a very uniform heating of the ore and consequently also a uniform chemical action.

Such a furnace will roast the ore with much more precision than a common furnace attended by such roasters as we can get in Nevada Territory, and it requires on that account less time. One man can attend several furnaces. The consumption of wood will be less; and no cooling at the working door can take place as in other furnaces where the door must be constantly open, whereby a great mass of cold air is drawn in, diminishing the draft at the fire place. The expenses are thus considerably reduced, but the most important advantage lies in the more perfect roasting which gives a better result in extracting the silver.

LOSS OF SILVER

In Roasting Different Silver Ores.

Sec. 37. In roasting the ore by the oxydation method without salt, the per centage of loss of silver will be higher, when there is a great deal of metallic silver in the ore, or when it is produced during the roasting, or when the mass of ore assumes a loose condition, admitting the air to permeate it. The loss increases also when the richness of the ore decreases, or when for some reason the temperature must be kept high. The sulphate of silver in contact with the oxyds of other metals suffers a greater loss than arsenate or antimony of silver, because the sulphate is more easily decomposed by oxyds at a high temperature and reduced to the metallic state. In roasting the ore without salt, silver is not only lost mechanically, being carried out with the draft, but chiefly chemically by the conversion of the metal into the oxyd of silver, which is volatile. If not combined with antimony, it deposits itself in a metallic form, for it leaves its affinity for oxygen at a lower temperature.

At Mansfeld, where Ziervogel's method is practiced, in extracting silver from copper matt the loss of silver, according to the recent accounts of Dr. Heinbeck, is 7·06 per cent. in roasting and 1·20 in extracting the silver, making the total loss of 8·26 per cent. This is considered a very flattering result. Taking into account the stuff collected from the dust chambers, the loss will be diminished somewhat.

In the chloridizing roasting, if properly conducted and if there are no base metals in the ore, the loss of silver is less than in the oxydizing roasting. However, the circumstances which determine the loss are different and numerous, but generally speaking the loss by roasting is between five and fifteen per cent. Losses above fifteen or below five per cent. are exceptions.

The base metal ore of the Ophir's northern claim on the Comstock lode, containing lead, zinc, iron, copper, and antimony, which I treated for the barrel amalgamation, mixing it with fifty per cent. of pure ore (the latter is now worked by itself), after having been carefully roasted, suffered a loss of between five and eight per cent. of silver. The accounts of the present losses in some works where roasting is going on are considered so low, that, allowing one or two per cent. for the amalgamation, there is hardly anything left for loss in roasting. This is evidently a mistake, founded on an improper mode of taking samples for that purpose.

AMALGAMATION OF ROASTED ORE.

a BARREL AMALGAMATION.

SEC. 38. The amalgamation in barrels is not adapted to ore containing gold. Unroasted ore has been tried with chemicals unsuccessfully. The construction of the barrels does nor differ much in the different works of Nevada Territory. They have a cylindrical shape, the diameter and depth being nearly equal. The staves are three to four inches thick. There are two sizes in use. The smallest capable of receiving from 1,000 to 1,300 pounds of ore, are thirty-two inches each way; the larger, receiving a ton of ore, measure from forty-four to forty-eight inches in the clear. The motion is imparted by cog-wheels although belts fitted directly on the barrel are preferable. The stoppage and starting by means of tightening pullies is more easily effected without jar.

Above each barrel is a wooden or sheet iron funnel, large enough to receive one charge. By means of a hose, fastened at the mouth of the funnel, the barrel is charged with the required quantity of ore in a short time. When this is done, from one hundred and sixty to two hundred pounds of wrought iron are introduced. The iron is in pieces of different shape and length, but pieces over five pounds, and such as have rough and sharp edges and ends, wear the staves too much. After the iron, cold water is added in such a proportion as to form a thickish paste. One ton of ore may require five hundred or six hundred pounds of water, according to the fineness of the ore.

The barrel is then closed tightly and set in motion, at the speed of about twelve revolutions per minute. After two hours' run, the barrels are stopped and examined. By this time the mass of ore should be of such consistency as to allow the forming of a soft ball with the hand. If the pulp adheres to the hand and fingers so that no ball can be formed, some thirty or fifty pounds of ore may be added, but if, on the contrary, the stuff is so dry that it crumbles into pieces, some more water must be added. After the ore has been found to be in the right condition, the barrels are charged with quicksilver. One thousand pounds of ore require five hundred pounds of quicksilver. The opening is closed again and secured by means of a screw. The motion must be changed now to eighteen or twenty revolutions per minute.

Four hours after the quicksilver has been charged, each barrel must be examined again. For this purpose the barrel is stopped and opened. A wooden stick, about an inch thick, is dipped into the pulp and withdrawn. If the ore is so diluted that it runs down from the stick, forming a long thread, the quicksilver and the iron sink to the bottom, the amalgamation is imperfect, and the iron does useless damage to the barrel. Some dry ore may be added, but after this time no more. If, on the other hand, the ore crumbles from the stick, or if it is so stiff that it does not adhere at all, the suspended quicksilver and iron have no chance to change their places, and the amalgamation will give a very poor result. An addition of water is therefore in this case necessary. The barrels are put in motion again and continued for fourteen hours, so that the period of amalgamation, from the introduction of quicksilver, will last about eighteen hours, after which the barrels are filled with water, set in motion at a reduced speed, and after one or two hours run, discharged.

The discharge is performed in different ways. Opposite the feedhole in the barrel is a small hole, shut by an inch screw. Through this hole the quicksilver is let out first, directly into the filter or into a common receiver. As soon as the mud appears, the screw is put in, the plug of the feedhole removed, the barrel turned over, and the tailings discharged into a large inclined trough below the barrels, leading into agitators, where the tailings are diluted with more water, in order to allow the settling of the particles of quicksilver and amalgam, requiring five or six hours, when the tailings are discharged. The contents of the barrels, quicksilver and tailings are also discharged at once into the agitators, the bottoms of which have a conical shape, like Wheeler's agitator. (See Sec. 51, Fig. 26.)

The agitator is a tub, five or six feet in diameter, and about the same in height. On the perpendicular centre-shaft are four arms with staves, three or four inches apart, performing the stirring. The shaft makes twelve revolutions per minute. In Washoe Valley there is an agitator, sixteen feet in diameter. This size can not be recommended. The purpose of the agitator is to have such a motion and such a dilution that, while the earthy particles are kept in suspension, the heavier but minute particles of quicksilver can sink by degrees to the bottom. If the mass is too thick, the quicksilver will be kept suspended; if it is too diluted, the sand settles with the quicksilver. The former result will also occur, if the motion is too fast, and the latter, if the motion is too slow. A sixteen feet agitator has a tremendous speed on the periphery when the motion near the centre is right. Even eight feet is too large, unless the agitator is not intended to effect a perfect separation, but is used like Wheeler's agitator. (See Sec. 27.)

As soon as the barrels are empty, and the iron pieces, which may happen to fall out, replaced, the charging is performed as before. The quicksilver, which came out of the barrels, is strained, and the amalgam retorted. (Sec. 42.)

The object of using metallic iron in the barrel is to decompose the sesquichloride, or chloride of iron, and to reduce it to protochloride. The chlorides of silver, copper, and lead, as well as some sulphates, after having been reduced to the metallic state, combine with the quicksilver, which is introduced after all these reactions have been effected by the metallic iron. In the absence of iron these chlorides would be decomposed by the quicksilver, which, forming sub-chloride of quicksilver, would decompose no longer, causing a great loss in mercury, and the amalgamation would be imperfect.

Metallic copper, in place of iron, acts with little less energy than the iron, but, not reducing the copper and lead chlorides to the metallic state, it renders a very pure amalgam. It is by no means necessary to use pure copper in the barrels. In treating the copperous silver ore from the Heintzelman mine in Arizona, I was obliged to procure copper by liquation of copperous lead. Being limited in regard to heat, for want of firebrick, the copper could not be refined properly, having a grayish red color, on account of some lead. This copper, as well as black copper, bought from Mexicans, gave a very favorable result as to the quality and quantity of the silver extracted.

b. Amalgamation in Dr. Veatch's Steam Tubs.

Sec. 39. The principle of this amalgamation, which is performed at the Central Works (Virginia City) is, in regard to chemical procedure, the same as that of the barrel amalgamation. Chlorides and sulphates are decomposed by wrought iron or copper plates. But, while the barrels discharge the quicksilver after each amalgamation of a new charge, in order to save it from destruction by decomposition, the steam tubs retain it for many charges, according to the richness of the ore. The quicksilver therefore takes an active part in decomposing the chlorides, forming calomel. In this combination the quicksilver is always lost. The steam has no chemical action, but it may influence the amalgamation by its temperatnre. In regard to the mechanical part of amalgamation, these tubs differ entirely from the barrel arrangement. There are wooden tubs about four feet deep and four feet in diameter. The bottom is made of cast iron with three circular openings for the reception of perforated plates, also of cast iron, below which are the steam chambers. The holes are very fine, about two inches apart. In the middle of the tub is an upright shaft, suspended on a box outside of the tub. There are three arms attached to it, each having three copper or iron plates hanging perpendicularly in concentric lines. The movable cover has an opening in connection with a flue by which the steam and some quicksilver are carried into cooling tanks.

The steam is forced through the perforated plates into the pulp, throwing the quicksilver in globules of all sizes constantly through the whole mass, causing a very perfect contact between the ore and mercury. The iron plates—or if the ore is very copperous, copper plates,—nine in number, have a circular motion, cut the ore with the edge, require very little power, and assist the motion of the pulp. The decomposition of chlorides goes on very rapidly in consequence of the heat and contact with the plates, which expose about 3,600 square inches of surface to a mass of six or eight hundred pounds of ore. Under such circumstances the amalgamation may be effected in five hours, but as a matter of course, the result depends entirely on a proper roasting. With regard to saving the gold in silver ore, this amalgamation may be equal to that of the barrels. Gold requires friction, silver chemical action. I could not ascertain the loss of quicksilver, but very likely it must be greater than in the barrel amalgamation, not only on account of being converted into calomel to some extent, but also by being dispersed by the force of steam, which should for that reason be carefully regulated. The mechanical loss and that by evaporation, in consequence of considerable heat in the tubs, is not important, as the greater part of the quicksilver is condensed in a cooling tank.

This amalgamation is superior to barrel amalgamation in regard to the time and the amount of power required, still there are some inconveniences which have not yet been removed. The choking of the fine holes in the perforated plates by amalgam, the cleaning of which is troublesome, might be avoided by discharging the quicksilver after each amalgamation; but then the plates would come too much in contact with the different salts of the ore, which doubtless would enlarge the holes and cause more damage than benefit. The cleaning of the tubs is injurious to the health of the workmen, if no time is allowed for cooling or precipitation of the vapor of the mercury.

The roasted and finely pulverized ore is spread on a platform moistened with water, and, after the plates have been set in motion, the quicksilver, water, and some steam introduced. The amalgamator is charged with six or eight pounds of ore by means of a shovel. If there is too much steam, the ore will be thrown up with the quicksilver against the cover, if on the other hand the steam has too little force, the amalgamation is delayed. The amalgam deposits partly on the plates, where it must be removed by iron scratchers without taking out the plates. After four or six hours, according to the richness of the ore, the amalgamators are discharged into agitators.

c. AMALGAMATION IN PANS.

SEC. 40. Ores containing such compounds of silver as cannot be treated satisfactorily in pans, or concentrated tailings of the pan amalgamation, which contain chiefly such silver ore as has resisted the action of chemicals, will give the best result after having been subjected to a chloridizing roasting. The presence of base metals will of course render the amalgam impure, but this depends partly on the greater or less attention given to the roasting. A certain amount of copper in the bar is not so injurious after all, increasing only the expense for transportation by express.

If the ore is of such a nature that a great deal of lead enters the amalgam, the metal must be refined, and in this case the calculation will show whether there is more economy in working the ore without roasting, even if more silver be lost. Generally, if in treating ore of about one hundred ounces per ton by means of roasting, not more than ten per cent. of silver were saved, it would not pay to roast. If the ore is very rich, especially when there is also gold in it, the proper way of reducing it, is to amalgamate it in pans without roasting, save the tailings in large tanks, roast them subsequently and work them over in pans, either alone or mixed with raw ore. In this case, although the water always carries off some ore in form of the finest slime, yet the loss is less, or may be a great deal less, than the loss in roasting some qualities of rich ore. We must take into consideration the fact that if the rich ore be treated by roasting previous to amalgamation, it must suffer not only the chemical loss but also that caused by the draft of the furnace; whereas the tailings, when exposed to the furnace after amalgamation, will have only the sixth part or so of the value of the original ore.

Ores containing antimony to excess, like that of the Sheba lode (in Humboldt County, N. T.), or combinations with zinc and iron, like the ore from the Rappahannock lode (Palmyra District, N. T.), cannot be treated in pans without roasting; also, auriferous arsenical or iron pyrites require roasting.

The amalgamation can be executed in different pans; but in order to save quicksilver from being changed into subchloride, the arrangement must be made that after each charge tailings and quicksilver be discharged. The construction of Wheeler's and Hepburn's pans makes such discharge necessary. On this account, and because they are superior in many respects, they are preferable to other pans. But the dies in Wheeler's pans form curved grooves in which a part of the quicksilver and amalgam remains after the discharge. They must be replaced by a bottom of one piece of hard iron or dies without space between them.

The ore is then introduced according to the capacity of the pan and the volume of the ore; for instance, five or six hundred pounds into one Wheeler's pan, or ten hundred pounds into Hepburn's, with a sufficient addition of water. The steam must be used moderately. The ore undergoes now the same process as in the barrels. After one hour's time, when the most of the chlorides and sulphates are decomposed and the silver reduced to the metallic state, sixty pounds of quicksilver are introduced. The amalgamation now goes on with the same speed and moderate temperature for two hours. Then all is discharged. The pan is charged again, and the procedure is the same as before. Wheeler's and Hepburn's pans alike require about three hours for the amalgamation of a charge. Others may require four or five hours, according to their serviceableness.

If the ore renders an impure amalgam, the impurity can be best ascertained after the first discharge by taking a sample of about ten grains of amalgam, which is heated to red heat by fire under draft, in order to get rid of quicksilver. This sample thus retorted must be examined under the blowpipe (see Sec. 11). If there are no other metals in it but iron, there is no remedy save better roasting. The chlorides of other base metals can be destroyed by using some quicklime, pulverized lime rock, or clean wood ashes. One or one and a half per cent. of lime is charged with the ore. After a quarter of an hour's time, a small portion of the ore is taken from the pan, put into a porcelain cup, and, by means of a piece of copper, stirred with some water and a few drops of quicksilver. If the quicksilver be covered with a black coat instantly, some more lime must be added. Fifteen minutes later another sample is taken in the same way, and so on till the quicksilver does not appear black, or very slightly so. Too much lime is injurious to the extraction of silver.

This method is only a temporary remedy. It can be better executed in the furnace. Such ore, introduced with about fifty per cent. of raw ore may prove a very good chemical.

III. COLD PROCESS.

AMERICAN HEAP AMALGAMATION OR PATIO.

Sec. 41. The patio amalgamation, where wood and water are scarce, the ore suitable, the climate favorable, and labor cheap, is a very good process, and in many cases cannot be replaced by a better one. The climate of Nevada Territory, at least in the summer, is very favorable for the patio; but several months of the winter time do not permit this process, without having proper buildings, in which, by the aid of steam and other arrangements, the required temperature is obtained, although the sunbeams and open air assist the process, independent of the temperature. On the other hand, there are the disadvantages that the expenses of steam power, horses, and men are comparatively much heavier here in Nevada Territory, than for instance in Mexico or South America. Besides, it is hardly possible that this process should justify its use, along side of pans working three or four tons of ore in twenty-four hours, with less expense and similar or better results, according to the nature of the ore.

Ores containing gold cannot be treated by patio, unless that metal is extracted first in some other way.

The best and at the same time cheapest method is to extract gold by the new improved pans, offering also the advantage of the finest grinding, required for the patio, but in this case the tailings of poorer ore would not pay to be worked over by the patio process.

The ores most suitable for this process are Brittle-silver ore, Polybasite, Rubysilver, Bromyrite, Iodyrite, Silverglance, and Chloride of Silver. The last two, on account of their toughness and ductility, amalgamate with little more difficulty than the other compounds.

Silver, combined with copper and antimony, must be slightly roasted before being subjected to patio amalgamation. It requires generally a very low, dark red heat, being stirred only at intervals for about twelve hours. Such ore can often be treated without magistral. If similar ore be taken into the patio without roasting, it remains cold, when the usual quantity of magistral is taken; and an abundance of it does not effect a proper extraction of silver. If on the other hand such ores are roasted too much, the patio appears very hot. It is well to let it rest after roasting for a few days in a moist condition.

Argentiferous zincblend, pyrites, and some other combinations cannot be treated by patio without a perfect roasting. Gold ores and argentiferous lead ores are entirely excluded from this process.

The patio process was tried in Virginia City, and is at present practiced in Washoe Valley and on Carson River. The theory of this process does not explain the chemical procedure satisfactorily. Salt and the magistral, containing about eighty per cent. of sulphate of copper, prepared from copper pyrites by roasting at a moderate temperature, generally with an addition of salt, are the two agents, acting and reacting mutually on each other and on the sulphurets in the presence of quicksilver. To these reactions the amalgamator (*azoguero*) gives his particular attention by making frequent tests with his horn spoon or other instrument. Notwithstanding all his care, it is sometimes found necessary to add to the mass a portion of quick lime, little or much, as the case may require, for the purpose of counteracting the injurious effects of too much magistral. The blue vitriol in place of magistral is less fit on account of its vigorous action upon the quicksilver.

The loss of quicksilver occurs here, as in other operations, for two reasons: the chemical, or "consume," and the mechanical, or "loss." The chemical is almost invariably equal to the weight of the silver extracted. The mechanical depends on the attention and ability of the amalgamator.

The ore is first ground very fine; then from thirty to sixty tons are laid out on the floor (patio) and mixed well with some water to a proper consistency. Samples are taken from all parts of the mass, a fire assay is made of them, and salt is introduced, amounting to two or four per cent. according to the quality of the salt and the quality of the ore. An excess of salt does not injure the operation. When the salt is well mixed with the mass, it should remain undisturbed for one or two days to permit the salt to be dissolved. The mass is then turned over and worked, by treading with horses or other means, to a uniform proper consistence.

The magistral is then incorporated, about one per cent., more or less, according to the quality of the ore, the temperature, and the situation. After the magistral has been well tramped in, the quicksilver is next scattered over it by straining through canvass. The amount of quicksilver first introduced is from one-half to two-thirds of the whole that may be required, which is about six pounds to one pound of silver, as ascertained by the assay of the sample.

After some days, the first part of the quicksilver has combined with enough silver to form a dry amalgam. When this is the case, one-half of the remainder of the quicksilver is added, mixed and trodden as before, and when after several days it is discovered that there is again dry amalgam, the last part of the mercury is introduced. Some days later, the amalgamation is finished, and then some more quicksilver may be added for the purpose of washing more easily.

The Torta (mass of the pulp) is tested twice a day by taking a small quantity of the pulp from different parts of the torta, and carefully washing it in a horn-spoon. The mud is washed away, leaving the amalgam on the bottom, also undecomposed sulphurets, and silver partially combined with minute particles of quicksilver (*limadura*). This *limadura* is separated from the other parts by a peculiar shaking of the spoon, and becomes the principal object for the azoguero's inspection. By its various states and color, the process of the patio is understood. There are, however, so many variations of appearance, that it requires a great deal of experience to form a correct judgment.

In the absence of magistral, for want of copper pyrites, blue vitriol or sulphate of copper is substituted. Mr. W. M. Brown, an experienced practical azoguero, who managed the patio operation at the Mexican mill, Virginia City, and afterwards on Carson River, prepares his magistral by roasting pan-tailings at a proper temperature, they containing a certain proportion of salt. These roasted tailings are mixed with the ore and act upon it as magistral.

RETORTING.

SEC. 42. The amalgam, a combination of gold and silver with mercury, must be separated, for the purpose of having the metal prepared for melting, also to regain the quicksilver for further use. The separation is called "retorting," a very simple process, being a mere distillation at a high heat. The quicksilver assumes a gaseous form, is led off through a cold pipe, and is condensed again to the metallic state, while the other metals remain in the retort.

The retorts used in Nevada Territory differ very much in shape and size. Those having an oval or cylindrical form are better than flat-bottomed retorts, which always require a separate trough for the amalgam, on account of the obstructive corners; besides, they have the disadvantage of being more liable to burst. The retort most used at present is the cylindrical retort. (Sec. 55, Figs. 20, 21.) The egg retort, of an oval shape, but lately introduced, seems to permit a more convenient charging and saving of fuel. This retort contains movable troughs or shelves for the amalgam.

Before charging, the lower part of the cylindrical retort must be coated with a soft pulp of fine, sifted wood ashes, by means of a long rod, to one end of which a rag is tied. The same precaution must be observed with the troughs for the egg retort. When dry, the amalgam is introduced into the retort in balls or pieces. Sufficient room at the top must be left for the vapors of quicksilver. If overcharged, some amalgam may come out at the pipe-hole. A retort of the size as described in Sec. 55 should not be charged with more than eight hundred or nine hundred and fifty pounds of silver amalgam.

When filled, the retort must be closed carefully, so that no vapors of quicksilver can escape anywhere. For this purpose the face of the door is covered with a paste of pure, fine, sifted ashes, about half an inch thick. The paste should be soft, but must not run. The correspondent face of the retort is moistened, and the door set in, fastened by the two wedges and tapped with a wooden mallet. No force at all is required. If the ash paste has the right consistency, the closing will be perfectly tight.

When this is done, the fire can be started immediately. As soon as the water, which cools the pipe, commences to boil, the fire must be kept very moderate for some hours, and increased when the boiling goes down. In about four hours after beginning, the retort should be dark red-hot, and kept in this state for two or three hours more, when the heat may be increased a little, not allowing a good light red heat, which would melt the silver partly, and also injure the retort.

When the pipe gets cool, and no quicksilver drops out, the retorting is finished. The water in the cooling pan, covering the quicksilver, must be clear. If it becomes milky, the distillation, on account of too much heat, was too sudden. It requires eight or ten hours to drive off all the quicksilver. A forced retorting does not gain much in time; the amalgam will generally be found not well retorted, and the retort will be used up in half the time. It is not the shape of a retort, but the time that is important for a good retorting. It is also a bad habit to open the retort before all is cooled down. A hot retort, even after the best retorting, contains mercurial vapors which are very injurious to the health, although the influence is not perceived immediately. When cold, the metal is taken out and broken into pieces for the purpose of melting.

MELTING OF RETORTED METAL.

Sec. 43. The retorted metal is of a spongy appearance, crumbling easily, and on that account is not fit for transportation or handling. To prepare it for transportation and to ascertain its value by assay, the retorted metal must be melted into bars.

The furnace is of simple construction (Sec. 56, Figs. 13, 14), square or round on the inside, and from ten to fourteen inches in the clear, according to the size of the crucible. The best are the black lead crucibles. Before using, they require to be annealed, that is, heated slowly by degrees to red heat. This may be performed in the same furnace on a moderate charcoal fire, or after melting, when the heat has gone down somewhat. The crucible must always be put first with the brim on the fire, so that the bottom is turned upwards. When the brim appears red-hot, the crucible can be turned without danger of cracking.

The most intense heat in the furnace is one and a half or two inches above the grate. The crucible, therefore, ought to stand always on a piece of fire-brick, so as not to be exposed to the cold draft of the air.

When the crucible is placed perpendicularly in the centre and covered, the charcoal is put around it and on and over the cover, some live coals on top, and the furnace is shut by the slides. When the crucible appears red-hot, the cover is removed carefully, that no charcoal falls into it, and so much silver as the crucible will hold (from twelve to twenty pounds) is introduced with a pair of longhandled tongs. A handful of borax is also added. The crucible is then covered, and the furnace, after a new charge of charcoal, closed again. In about half an hour the silver is melted down so far that another portion of silver can be introduced, and so on, till fifty or sixty pounds, which is about the right amount for a proper bar, are melted down.

The borax must be used in proportion to the impurities of the metal. If the retorted metal is white and clear, one handful is sufficient; but if the amalgam looks black, double the quantity is required. The borax dissolves the oxyds of base metals and earthy matters, which may happen to be in the amalgam, and forms a black slag, which covers the metal. Some amalgamations furnish amalgam, containing a great deal of sulphurets. When melted, the sulphur takes up so much silver as to form a combination similar to the silver-glance, containing eighty per cent. of silver. If there is metallic iron in the amalgam or crucible, it combines with the sulphur, setting the silver free. This combination of sulphur and metals, called "matt," is poorer in silver, in proportion to the amount of metallic iron present. The matt is more liquid than the slag, but heavier. It lies on the metal, below the slag. If rich in silver, the matt is tough, bluish gray; if poor, it is brittle and has a yellowish gray appearance.

After the last charge is melted, the cover is removed and metal and slag stirred well by means of a red-hot iron rod, then covered, and a good heat effected by another charge of charcoal, when the slag is taken off. This is generally done by a skimmer (Fig. 16), made of half inch round iron, rolled into a coil at one end. With this skimmer the slag is touched carefully, not dipping too deep. The slag adheres to the skimmer, which is taken out immediately after it comes into contact, and tapped slightly against the stone floor, or a wet board. With this cooled coating the **skimmer** is dipped again into the **crucible,** and the adhering slag **treated** as before. This is **done** repeatedly, till **nearly all the slag** is removed, when by a smaller **and lighter** skimmer the surface of the metal **is cleaned entirely in the same way.** It has often been **mentioned that when** metallic silver is exposed in a **melted** condition **to the air,** oxydation takes place and causes a loss by volatilization. **By adding the borax with the first** charge, the silver will be covered **by it and protected;** but when, by skimming, all the **slag has been removed, and** the metal surface appears clear, **a handful of charcoal powder must be** thrown into **the crucible immediately after** the last slag has been taken out. **The crucible is then** covered, and the last heat applied.

If there be matt present, it, being more fluid **than the** slag, does not adhere so well to the skimmer. **After** the accumulated slag has been knocked **off from the** skimmer, it is dipped **as before, but must be taken out** quicker and the matt shaken **off by a** quick shake over a pan or **water tub. This operation** continues till all the matt is out. **When the metal** appears clear, a small piece **of** borax **is introduced, and** when melted removed by means of **a small** skimmer. Charcoal dust is **then** introduced and **a last** good heat given. **The crucible is** taken out by a pair of strong crucible **tongs (Fig. 15).**

It is placed before the mould, and the contents poured out in a uniform stream and not too slow. The cast-iron mould must be made hot, and if required, smoked over burning rosin, so that the whole inside appears covered with soot. The mould must stand level. As soon as the metal is in the mould it is covered immediately with charcoal powder.

In this simple proceeding, the **look of the bar** depends entirely on the greater or less purity of the metal. It will however answer the purpose, as there is no necessity to spend time and fuel in acquiring a nice bar, unless the bar has to be stamped with the value, ready for the market. If not stamped by a responsible firm, the bar must be remelted for this purpose, no matter how it looks. Hence there is no need for the millman to trouble himself in making a nice bar. If a handsome bar is desired, it must be observed that after slagging, the metal should appear with a smooth, mirror-like surface, so that objects may be reflected by it. If not, if the metal when melted continues to evolve impurities which cloud the surface, a new portion of borax must be added and stirred with a red-hot iron, or a red-hot slip of black lead crucible. Then the heat is raised again, and the operation may be repeated two or three times, always removing the slag before adding borax, till the metal appears lustrous.

But in many instances, especially when a great deal of sulphurets come in the amalgam, sulphur and other impurities are so abundant that it would take half a day or more to get rid of it merely with borax. In such a case, if there are about sixty pounds of metal in the crucible, after all slag and matt have been removed, four or five ounces of lead are introduced with the addition of some borax. The metal is then mixed well, the crucible covered, and the heat increased. When the charcoal has burned down to the brim of the crucible, the cover is taken off, and the slide doors of the furnace shut partly, so that only three to four inches of opening is left, otherwise the heat is oppressive to the workman. The metal must be stirred over, and the borax which takes up the oxyd of lead and other impurities, separated by the aid of lead, skimmed off and replaced by another piece of borax. The oxyd of lead can be seen distinctly in numerous little spots adjoining the borax. When the last particles of lead separate from the silver, the surface of the metal will brighten for a few seconds, assuming by degrees a clear lustrous appearance. When the borax is saturated by the litharge, the crucible will be attacked. It is therefore necessary to skim often and to add more borax.

Such metal, also amalgam from roasted ore, especially if obtained in large quantities, is not only cheaper and more economical to refine in a refining furnace, like the cupel furnace (Sec. 58) which should be of a smaller size, but even the melting into bars is more advantageous and a great deal less troublesome than the crucible melting, unless it is very pure metal. Silver, exposed to the draft in a melting condition, suffers a loss which increases with time and heat. For this reason the melting in crucibles is preferable, if the metal is pure, but when oxydation of base metals is required, in order to refine the silver, a refining furnace must be used.

A new test furnace ought to be dried by a slow fire, at least two days before the heat can be increased. When the test appears light red hot, a small piece of retorted amalgam is introduced in the middle of the test, and the door shut. If it melts into a bright silver button the amalgam can be charged as carefully as possible, otherwise the test might be injured, especially the first charge in a new test. Several handfuls of charcoal dust are introduced, the door closed, and a strong heat applied. The silver will soon commence to melt, making room for another charge. When the test crucible is full, containing sixty, one hundred, two hundred, or two thousand pounds, according to the size, good heat must be kept up for about half an hour, then stirred with a red-hot iron hook, the end of which is bent upwards so as not to tear the crucible. The charcoal dust will soon burn off, and must be replaced by another charge when the metal appears bright and clear. It is then ready to be dipped up and poured into moulds previously warmed.

But if dry, ash-like impurities appear on the surface of the metal bath some litharge may be thrown on it. This will fuse, and be drawn into the mass, taking the impurities with it. The metal must be stirred several times and a strong heat kept up, till after an hour or two, the surface of the silver will appear like a mirror without agitation. If, however, the silver should contain a good deal of foreign matter, twenty or twenty-five ounces of lead may be introduced. The metal will commence immediately to work, minute spots of litharge will arise on the surface increasing and gliding towards the test, by which they will be taken up. This will continue for some time, when the whole bath will be perceived to be covered suddenly with a bright cloud which disappears in a few minutes, leaving the metal clear. As a matter of course, no lead is necessary if the amalgam is already alloyed with it, which is easily discovered by the blowpipe (Sec. 2).

The hot mould is placed close to the door, and the silver poured into it by means of an iron ladle about six inches in diameter and two inches deep. This ladle must be made red hot before using. Near the close of the operation the heat must be increased and kept on till all the silver is out. A small quantity which cannot be dipped out always remains. This is left till it cools and becomes hard, when it is easily removed by an iron rod. This, however, must be done without delay as soon as the cake gets hard; otherwise, if too late, the test would certainly get damaged. This is of course not necessary, if another charge of silver has to be melted.

All the slag and matt must be carefully gathered till a convenient time is found to remelt it. It happens sometimes that the skimmer takes up a few drops of metal with the slag, especially if there is not sufficient heat in the crucible. The matt always contains more or less silver. The slag and matt, when they are to be melted, are broken into small pieces by a hammer and mixed with ten per cent. of soda-ash or soda. An old crucible of a good size (No. 50.) which appears strong enough to stand the melting is placed in the furnace in the same way as for melting retorted amalgam, and the mixture is introduced by means of a scoop, filling two-thirds of the crucible. It requires about one hour to melt one charge. When the fusion is complete, the contents of the crucible is stirred first with a rod of iron a quarter of an inch thick which is bent over like a hook. When after some time, the rod gets red-hot and the hook melts down, some old nails or scrap iron must be added and stirred with a thicker iron rod from time to time. Another test is made with the quarter-inch iron rod, and when that rod no longer assumes a white heat at which it melts off, a last strong heat is applied and the slag is removed by an iron ladle which is cooled in water after each dip. When about two-thirds has been removed in this way, the crucible is charged again and managed as before.

Silver matt when thus treated transfers its sulphur to the iron, and the silver is reduced to the metallic state, accumulating at the bottom of the crucible. So long as the test iron melts off in a short time, more iron must be added. When, after four or five charges from which the slag has always been skimmed off, sufficient metal accumulates in the crucible, it is taken out and the contents poured into a warm mould. By adding lead, the extraction of silver is more perfect, but then it requires cupellation (Sec. 46.)

The borax slag, especially when matt is present, contains very little silver, except that slag which involves the matt in small globules. Therefore the shortest way of beneficiating slag and matt, is to throw it on the ore before the battery and to treat it in pans with the ore, provided that no roasting is in progress, because the slag particles melt at a very low heat, thus forming hard little half-melted lumps, which however are not injurious if pan amalgamation is adopted.

ASSAY OF THE BAR.

SEC. 44. It is not the intention to describe here the way of assaying for the purpose of stamping the value on the bar, as required for the market, but merely to ascertain the value for transportation, or to have some check on the value to be ascertained in the assay office where it is remelted.

After the bar has been cleaned and dried, if cooled in water, and that is generally done to save time, two little pieces are cut off from the corners, one on each flat side, and the bar weighed on a good platform-scale. The two pieces are flattened on an anvil thin enough to be cut with a pair of scissors, and heated to a red heat on a piece of charcoal with a blowpipe.

As the upper and lower sides of the bar do not differ

much in regard to fineness, provided that the melted silver was well mixed at the moment before it was poured into the mould, it is sufficient to weigh on the assay balance about five grains of one piece and to add as much of the other, so as to weigh very exactly ten grains in all. It is understood, of course, that the silver chips must be perfectly clean. These ten grains are wrapped in a little piece of sheet lead of about ten grains weight, or more if copper is in the metal, and introduced into a cupel by a pair of cupel-tongs. The cupel must be light red-hot before the assay is introduced. It is then cupelled according to the process described in Sec. 19, or the cupellation can be performed by the blowpipe (Sec. 12). The silver button is taken from the cupel with a pair of pincers and cleansed of the adhering particles of bone-ash with a toothbrush. It is then hammered flat on an anvil and annealed again with the blowpipe or in the muffle. It is now weighed on the balance and the weight noted. After this, the amount of gold must be ascertained by parting it from silver. For this purpose, pure nitric acid (about three-fourths of an ounce) is poured into a glass matrass or tube and the silver plate dissolved as described in Sec. 19. The weight of the gold is also noted, and the calculation made as illustrated in the following example:

For instance, the bar weighed on the platform scale forty-one and one-half pounds. This must be multiplied by 14·58 to find the amount of ounces. The silver button after cupellation weighed, say, 887 and the gold 131. The value of the bar will be found thus:

The bar weighs 41·5 × 14·58 = 605·07 ounces.
The fineness in silver is 887−131 = 756 ×1·30 = value per oz. $0,98·2
" " gold............ 131 × 20·67 " " 2,70·7

Value of gold and silver per ounce............... $3,68·9
Hence the value of the bar is 605·07 × 3·689 = $2232·10.

If the same kind of ore is treated by a permanent process, the fineness of the silver varies generally very little in regard to base metals, except when it comes from the roasting process. It is then, if not required to know the very exact value, sufficient to take the average fineness of some bars, say 887, and to make the assay only on gold, the amount of which should be subtracted from 887, leaving thus the silver for the calculation very near the real amount.

If, on account of a high amount of gold, the assay should not dissolve in nitric acid, pure silver must be added. (Sec. 19.)

IV. MELTING PROCESS.

Sec. 45. The melting of silver ore, carried on in several metallurgical works of San Francisco, other parts of California, and in Pleasant Valley, N. T., is comparatively an expensive method. It is not very likely that silver ore ever will be worked advantageously by melting, unless the amount of lead or some particular circumstances should decide for it.

Very rich ore, which allows the use of sufficient fluxes, may justify a proper melting; but even then, amalgamation in pans without roasting, with a saving of the tailings to be worked over in pans or barrels after a chlorodizing roasting, will be preferable to the expensive melting, which requires experienced and skillful hands, and a considerable investment of capital.

The reverberatory furnaces, which are used in California or Nevada Territory, have nothing peculiar in their construction, and resemble, generally in arrangement, shape, and dimensions those described in metallurgical books. These furnaces require the best fuel, either imported coal or artificially dried wood, and do not seem to be suitable for the circumstances of California and Nevada Territory. This, of course, does not allude to ore like pure galena, which does not require such a heat as the silver ore, the latter being generally accompanied by various foreign substances. However, a description will be given of a crucible furnace which was planned by the writer for melting the rich Ophir ore in San Francisco, and which answered the purpose perfectly.

The blast furnaces, where the burning of coke, anthracite, or charcoal, is forced by compressed air, concentrated in a small space, thus affording a high temperature, even if a lower quality of fuel must be used, are more suitable, under our circumstances, for melting silver ore, provided that sufficient lead ore, containing at least fifty per cent. of lead, can be mixed with it.

This kind of furnace is also more easily constructed, not absolutely requiring fire brick, for which some kinds of sand stone, conglomerate with predominant quartz, or some clay slate may be substituted. The disadvantage is the want of power to drive the bellows, which is the soul of the furnace. The shape and dimensions are described in all metallurgical books. Besides, the dimensions and the whole arrangement should be accommodated to the locality and the ore.

The crucible furnace (Sec. 57, Figs. 32, 33, 34), derived from the Mexican cupel furnace, can melt, according to the size and quality of ore and fuel, from half a ton to three tons in twenty-four hours. The ore is pulverized and sifted in a sieve of about six hundred and twenty-five holes to the square inch. The ore can be melted directly, when mixed with fluxes; but it is more economical to roast it after it has been pulverized, in order to drive out as much sulphur as is possible, at a low heat. If subjected to melting without roasting, a great deal of metallic iron is required to consume all the sulphur.

Iron, lead, silver, and copper, combine with the sulphur and form a matt which covers the lead below the slag. This matt must be roasted and treated like ore. It will also be formed, to some extent, even after a good roasting, and it would not be advisable to roast so perfectly as to prevent the formation of matt, because the slag will always be poorer in silver, if some matt is made in melting.

The **roasting is** performed in a common roasting furnace, at a **very low** heat, with diligent stirring, continuing **till** no sulphurous gas is emitted, which can **be ascertained** by the odor. This operation may require three hours. To each **new charge** of **ore** some pulverized matt should be **added, so that it shall** not accumulate. It may **be** pulverized with the ore, or separately.

The roasted **ore is then mixed with** fluxes, of which there **are three classes:**

1. **Fluxes, dissolving the** different earths and oxyds of base **metals.** Such fluxes **are:** soda ash, litharge, slag, and **silicia or quartz.**

2. **Fluxes,** decomposing sulphurets: metallic iron, litharge, and lime; also, soda ash.

3. **Fluxes,** collecting the gold and silver: lead, in the the shape **of** granulated metallic **lead, or as lead** ore; litharge, or hearth (the mass from the refining furnace).

The litharge, or lead-oxyd, which is obtained from the cupel **furnace, consisting of 92·8 lead and 7·17** oxygen, eagerly dissolves **the quartz and other** earths forming silicates. It is also a **powerful** agent in decomposing sulphurets by its oxygen, **creating** sulphurous acid and **metallic** lead, which latter combines with the **metal of the sulphuret** and all metallic particles with which it may come **in** contact.

The **addition** of slag in melting has a double purpose: first, to dissolve **the** earths and oxyds of metals; second, to regain the **particles of** lead and matt which were **drawn out** with the slag, **in cleaning** the surface of **the** lead before tapping. If there be not a sufficiency of such slag, poorer slag must be added.

The silicia, in the shape of sand, dissolves the oxyd of iron. The addition of sand is required when roasted matt is melted without sufficient ore to mix with it. In this case, if no sand were used, the oxyds of iron and lead would attack the mass of the hearth considerably.

The lime combining with the slag, protects the oxyd of lead against such combination and decomposes the sulphurets indirectly.

The hearth of the cupel is used on account of being saturated with litharge.

The charcoal in the pulverized state decomposes the oxyd of lead and in part the sulphates. It is used in such proportion as to reduce one part of the litharge, while another part decomposes the sulphurets which may remain in the ore after roasting.

Metallic iron (granulated or borings) is the best agent for the decomposition of sulphurets forming iron matt.

Metallic lead takes up the gold and silver whenever it comes in contact with these metals or melted sulphurets. No melting of silver ores can be performed without lead or materials which produce lead. If there is no lead ore, granulated lead or litharge will answer even better (but not in a blast furnace). Litharge must be pulverized and sifted. The lead requires to be finely granulated. The granulation is effected in a wooden box (Fig. 19). The bottom and sides must be smooth, and the box tight. The box hangs on a rope so as to allow a swinging motion. Several hundred pounds of lead are melted in an iron kettle and kept in a fused condition by a slight fire underneath. The temperature of the lead is right when a wooden stick thrust into it turns brown without causing a boiling motion. About twenty pounds are then taken with a hot crucible or an iron ladle and introduced into the box, in which a handful of pulverized charcoal was previously put. This must be repeated at each charge.

The workman immediately takes hold of the handles, a, and swings the box in such a way that the lead slides from the side, b, over to c, and back, and so on repeatedly. When the lead becomes more compact by cooling, the swinging must be shorter and quicker, so that the lead strikes the sides with force till it falls in dust. The box is turned over and another charge granulated. The lead thus granulated must be sifted in a sieve of about six hundred and twenty-five holes to the square inch.

Lead ore is preferable to lead or litharge only in case that it contains some silver, or if it is cheaper. The ore must be selected carefully and all rubbish separated. It should contain seventy or seventy-five per cent. of lead, never less than fifty. If the ore is galena (sulphuret of lead) it should be mixed with the ore in the right proportion before pulverization and roasted together with an addition of matt, obtained in course of melting. The proportion of ore and fluxes may be changed according to the nature of the ore; the following mixtures, however, may be given:

No. 1. MIXTURE FOR ORES WITHOUT ROASTING AND BEFORE LITHARGE IS OBTAINED FROM THE MANIPULATION.

Silver ore................................100 pounds.
Granulated lead (or 200 pounds lead ore).......... 85 pounds.
Soda ash....................................... 25 pounds.
Iron... 25 pounds.
Lime (and 25 pounds slag when obtained).......... 3 pounds.

No. 2. MIXTURE FOR ROASTED ORE.

Silver ore................................100 pounds.
Granulated lead............................... 85 pounds.
Soda ash...................................... 20 pounds.
Iron.. 8 pounds.
Lime.. 3 pounds.

No. 3. MIXTURE OF SILVER ORE AFTER PRODUCTS OF SMELTING ARE AT HAND.

Silver ore................................100 pounds.
Granulated lead............................... 25 pounds.
Litharge...................................... 75 pounds.
Hearth.. 10 pounds.
Soda ash...................................... 15 pounds.
Charcoal...................................... 5 pounds.
Iron.. 8 pounds.
Lime.. 3 pounds.
Slag.. 25 pounds.

When the melting is executed with an addition of lead ore, No. 2. and No. 3 do not require granulated lead, but it is always very useful to add sufficient litharge. The quantity of flux required depends much on the quality of ore. If the slag is too thin, it is not necessary to use so much soda ash.

When the furnace is white-hot, the mixture is introduced by means of a scoop or shovel at the flue door (Fig. 32, i), and spread to about half the length of the flue, between the door and the crucible. The ore may lie five or six inches deep. The door is closed, and the firing continued so that the flame reaches the end of the flue. The ore soon commences to melt, and runs into the crucible. A new charge must be introduced as often as the melting ore makes room for it. On the melting surface of the ore innumerable lead globules arise, which, taking up silver, grow bigger by joining other globules, and roll down into the crucible, followed by the slag. The lead separates in the crucible from the slag and matt, and it is very important to open the front door, k, often, in order to mix the slag and lead well by stirring with an iron rod, which is bent hook-like in such a way that the crucible may not be injured. The more the lead is brought into contact with slag, the poorer the latter will be in silver. The melting goes on till the crucible is nearly full; but care must be taken to stop charging in time, so that, when the crucible is full, no ore shall be in the flue. The melted mass is now stirred again, the door closed, and a good heat applied for fifteen minutes, after which the firing is stopped and the slag channel opened by means of a bar.

The slag runs out in an iron car. The channel is then shut again by a paste, formed of one part of ashes in bulk, and one of loam, well mixed. The paste must be made soft, but still so that it may be formed into round, long pieces, which, after being dipped in water, are laid directly into the hot channel and pressed with a piece of red-hot flat iron. This operation must be repeated after each introduction of the paste till the channel is filled level with the rim of the crucible.

The fire is started again, the flue charged with ore, and the melting executed in the same way as before. After the slag has been discharged five or six times, it may be examined by an iron ladle, to ascertain how much lead or matt has accumulated in the crucible, and when it is found that the matt is only three inches below the bottom of the slag channel, the slag is first discharged into the car, and when it has run out, the remainder must be drawn off by an iron hoe through the channel till the matt appears free from slag. The slag thus drawn off is mixed with the ore again, as mentioned before, for some matt will unavoidably be drawn out likewise. When all the slag has been removed from the crucible, the lead and matt must be tapped into the open hearth, B'. The taphole is opened with a chisel-like pointed bar, boring and picking the loam mixture in the hole, not using too much force, till the lead commences to run out.

When all is out, the taphole must be closed again without delay. For this purpose a piece of charcoal, about two and a half inches long, pointed at one end, wrapped with loam at the other, and shaped like the taphole, is fastened to a rod, and introduced so that the point of the charcoal shuts up the hole in the crucible. Two or three other loam balls are applied on it, tapping each with a wooden rod.

The matt, separated from the lead, hardens quicker than the lead below it in the lower hearth, and can be removed in one piece with an iron fork. The lead is then dipped up with ladles and poured into warm iron moulds. The lead bars weigh twenty-five or thirty pounds, a convenient size for handling at the cupel furnace.

All the matt is turned into the ore and pulverized with it. In roasting, the sulphur is burnt off and the iron oxydized, being thus a good flux for the silicia, while the silver is absorbed by the lead.

SEPARATION OF LEAD AND SILVER, OR CUPELLATION.

SEC. 46. The lead, obtained from melting silver ores, must be separated from the silver by an oxydizing process, called cupellation. To this the lead is subjected either directly, or, if very poor in silver, after concentration in a smaller quantity of lead, in order to reduce the expense of cupellation. This (Pattinson's) concentration has been introduced by Capt. Mead in Pleasant Valley, N. T., the result of which is not yet known.

Some general remarks on the concentration by crystallization will follow, after the cupellation shall have been described.

The cupel furnace (Sec. 58, Figs. 35, 36), after having been perfectly dried by a slow fire for two or three days, must be made light red-hot, before the lead can be introduced. A piece of lead is placed on the hearth, and observed. First it will melt and become covered with an oxyd. If the heat is too low, the oxyd coat grows thicker, and the lead remains dark red. In this case the door is shut and the temperature raised by better firing. The oxyd crust will then melt and disappear, leaving a round, bright lead button. When this is perceived, two lead bars are introduced, placed on the fire-tile, and replaced by others when melted down, till the hearth is full. If the lead is not clean, there remains on the tile a kind of skeleton, mostly of matt, which is pushed into the test before another lead bar is introduced. The doors are closed, and the heat raised. After half an hour's time the surface of the leadbath is covered with a kind of slag, the scrapings, consisting of oxyds of lead and iron, and other foreign matters. These scrapings, to be melted, require more heat than the litharge, and are called froth. This froth is drawn off, over the litharge bridge (Fig. 36, *f*), in which a small canal, *c'*, is made just before drawing. For this purpose a piece of flat iron is prepared, about one and a half inches wide by one-half inch thick and five feet long, one end of which is sharpened and stretched a little, so that it assumes the shape of a chisel. With this instrument the canal is made by careful picking and scratching, about half an inch deep.

Over this canal the scrapings must flow from the furnace. To effect this, the slag is drawn first with a round iron rod over the channel repeatedly, holding the rod light on the surface of the bath, or otherwise some lead might escape. The scrapings, assisted in this way, will soon flow by themselves, but from time to time they require again the aid of the rod. The easy run depends also on the temperature; if too low, the scrapings will freeze and shut up the canal. This slag looks black, is heavy and brittle and shows a glassy surface. By degrees it changes color, becoming more yellow, flowing more easily, and assuming a scaly appearance. The litharge is formed now, and flows freely through the canal, which however must be constantly attended to by widening in case of need, or, by arresting the stream by putting a small lump of the hearth mass in the canal, if any drops of lead escape. The heat must be kept moderate and regulated after the appearance of the litharge. As long as the scrapings continue to run, the temperature requires to be higher so that the scrapings flow down to the floor. When so much of this slag is removed that the lead is exposed in the middle of the bath, the blast is introduced, first gently blowing, but it may be increased when litharge arises to such a degree that the lead is slightly moved by the wind and the litharge is driven against the sides. The lead bath brightens and must be always lighter in color than the hearth at the sides. When the litharge solidifies on the bridge or shortly below it, the operation requires more heat. The right temperature is then when the litharge flows to the floor, but when it flows on the floor in a red-hot condition, there is too much heat in the oven, causing a greater loss in lead by volatilization.

When the litharge ceases to flow at the right temperature, the canal must not be made deeper, but another lead bar placed on the tile. It requires much attention to keep the canal as uniform as possible about half an inch wide. This canal will be cut deeper gradually by the litharge. If the temperature is too high the litharge attacks the hearth-mass more, the canal will grow deeper and some lead may escape. When, therefore, after a longer use, or with excessive heat, the canal appears too deep, it must be filled with hearth-mass and another canal opened alongside of it as flat as possible, just to permit the flowing of the litharge. The charging of lead bars must continue without interruption, regulated by the canal and the formation of litharge as above described.

In drawing the litharge, it is an error to let all flow out. Particular attention must be paid to have at least four or five or even more inches of the surface of the lead always covered with litharge, which forms a flat ring on the periphery of the bath. The longer the litharge is in contact with the lead the poorer it is in silver. Another sign for the right temperature is in the lead fumes. If the heat is too strong, the lead fumes will arise in such a quantity that it is difficult to see the bath, which interferes very much with the drawing of the litharge. If on the other hand the temperature is too low, the lead vapors will disappear almost entirely, the litharge does not flow right, cools in the channel and stops. The right temperature is indicated by a moderate amount of lead vapors, so that most of the litharge can be seen distinctly and the flow is free, over the whole litharge bridge.

When after twelve or fifteen hours the whole surface of the bridge has been cut for canals, generally five or six of them, the first one must be opened again or better a new one made, but of course deeper than the first row. The lead bath will stand accordingly also deeper in the test. We now proceed in the same way, making new canals close by, when the first is getting too deep and so on. After from thirty to forty-eight hours of uninterrupted cupellation, the charging of lead bars is stopped. From this moment the last canal must be made deeper as the drawing of the litharge requires it.

During the whole process the formation of litharge can be observed, and it flows constantly on the convex surface of the bath towards the sides joining the ring of litharge, kept purposely to some extent, in large furnaces even twelve inches wide. At the end of the operation when the silver is concentrated and but little lead in it, the litharge appears less, but in larger spots, till at last a shining, playing veil appears, which soon covers all the metal. The blast must be moderated now and the heat increased. After some time the bath clears up again, emitting some litharge in minute particles which continue under strong heat for a short time. The blast must be reduced now to the lowest degree and then stopped entirely, when the silver commences to play rainbow colors, but the fire must be kept on. The silver is stirred with a red-hot iron rod once or twice till finally the metal appears bright and clear like a mirror. If there are eighty pounds or more of metal in the test, the silver may be dipped out with an iron ladle into hot moulds. Smaller quantities are cooled by the blast, also some cold water may be poured on the middle of the silver cake, yet only a little at a time, so as not to cool the hearth too much. When the silver hardens, it emits oxygen, forming figures or towers of different shape. The silver is then tried with an iron bar, slightly knocking on it. If it sounds like solid metal, a pointed bar will lift the cake by introducing the point under the edge of the slab. The moment when the silver becomes hard, must be watched in this operation, else, if too late, the test might be considerably damaged. In case the cake is larger than the opening of the litharge bridge, it must be taken out through the fire-place.

The cupellation, executed in the described way, yields very fine silver, which need not be refined. Generally the cupellation is interrupted when the brightening occurs. In this case, however, the silver is not quite pure and must be refined in a refining furnace.

As soon as the silver is taken out, the furnace must be immediately prepared for another operation, if sufficient pig lead is at hand. For this purpose, three or four flat iron bars, each five feet long, are introduced into the furnace, so that one end of each may get red-hot, while the other ends remain cold enough to be handled. Meanwhile the same composition, of which the hearth was made (Sec. 58), is prepared with some water to a soft paste, well worked, and formed into cylinder-like pieces, about eight inches long and three or four inches thick, of which from four to six pieces may be required.

The canal is cleaned of litharge, and one of the prepared oblong pieces dipped into water, and by means of a flat, cold iron, placed in the hot canal. The mass introduced must be pressed and slightly beaten with one of the red-hot bars till it seems to be well united. This operation must be performed quickly, so that the oblong mass is quite wet when touching the canal, and is then instantly pressed and beaten. Another slab follows now exactly in the same way, and so on, until the canal is filled level with the hearth, forming a solid mass. When this is done, fuel is introduced into the furnace quite moderately for half an hour, in order to dry the new litharge bridge. The heat is then increased, and when the furnace appears light red-hot, the charging and cupelling of the lead is performed in the way described.

This operation can continue for one or two weeks,

according to the quality of the hearth. If, however, it were required to suspend the operation for one or more days, the hearth would crack in cooling, if not provided for. In this case, after the new litharge bridge has been made, the hearth and bridges are covered with ashes, and all draft shut as off close as possible.

In mending the canal, much attention must be paid to the right consistency of the mass, and to having it wet immediately before the application. The mass must be soft, but must retain the form given to it. If too hard, or if the dipping into water be neglected, the lead would find its way through, coming forth in drops. In such cases where the drops are perceived, the hearth is hollowed about three inches deep, and stopped with some of the mass, using a red-hot iron.

The litharge, if required, can be easily reduced to metallic lead by having a cylinder of cast iron in front of the litharge bridge. The cylinder is hollow, open at both ends, and the sides have a number of inch holes to admit the air. It stands over a cast-iron bowl, or on the floor, in which a hole is made for the reception of lead. The cylinder is filled with charcoal and set on fire, being about four feet high and twelve inches in diameter. The litharge is led over an iron plate into the centre of the cylinder. The oxyd of lead, fused and red-hot, coming into contact with glowing coals, is reduced to lead, and accumulates in the crucible below the cylinder, which must always be kept filled with charcoal.

REFINING OF SILVER.

SEC. 47. The refining of silver is the continuation of cupellation, when, as before described, the process is considered finished with the brightening of the silver. To this process all impure silver is subjected, but when copper or impurities are considerable, so that a proportionate quantity of lead must be added, requiring also a blast to effect the oxydation, and a litharge canal to draw off the litharge, then the refining is performed by way of cupellation. Retorted amalgam, obtained by the pan amalgamation without roasting, when melted into bars, is generally between 987 and 997 fine, not requiring any further refining. But in some cases the amalgam contains sulphurets, the sulphur of which remains to a considerable amount in the silver, when melted in the crucible. In pouring the metal into the mould, sulphurous acid is emitted, and its evolution continues till the metal hardens, causing a dull, uneven surface. Such amalgam, as well as that from roasted ore, or silver, generally, if obtained in large quantities, can be melted to advantage in refining furnaces.

The refining is performed in different ways. It is done in crucibles, having thus the smallest loss in silver; but only such silver as contains a small amount of impurities can be treated for this purpose in crucibles, which, on the other hand, besides requiring more fuel, are more dangerous than the hearths, on account of being liable to break, and also expensive. The refining is also executed in cast-iron dishes, which are lined with fire-proof material. These tests are placed in the refining furnace, exposed to the flame, or under a muffle, the last method requiring more fuel and silver not too impure.

The most proper way is to refine in hearth furnaces. (Sec. 59.) There is little difference between them and the cupelling furnaces in regard to size, and no blast is used in either. The consumption of fuel is moderate, and also the loss of silver, if properly attended to. The quantity of silver which can be refined at a time varies, according to the size of the furnace, from one hundred to two thousand pounds. These furnaces may be constructed to be heated either by flame or gas. The gas is produced by charcoal, coke, or anthracite.

The silver is introduced when the test appears whitehot. The door is closed and the fire kept up till all the silver is melted. The retorted amalgam, on account of its bulk, can be charged at once. Another portion is introduced and so on, till the test is full. When the last charge is melted, the silver must be stirred with an iron rod for a short time, and then the door is closed again for half an hour. If there are still ash-like impurities, swimming on the surface, they may be skimmed off, or one per cent. of lead is introduced into the silver and stirred again. The litharge will appear soon on the surface and, dissolving the dry scrapings, draw into the test. The stirring, at interval of half an hour may be repeated three or four times during the process. The silver becomes gradually clearer and brighter till finally the roof of the furnace is reflected on the lustrous surface of the bath. At this point the silver possesses the required fineness of about 999·5, and to prevent its volatilization its surface must be covered with charcoal powder immediately. The metal may be dipped up and poured into moulds, keeping up continually a strong heat, or it may be cooled in the furnace and taken out in the shape of a cake, in the same way as described in Sec. 46. The melting may require two or three hours and about the same time must be spent for refining, so that the whole process takes from five to eight hours, according to the quantity of silver introduced.

PATTINSON'S CRYSTALLIZATION PROCESS.

Sec. 48. This process, for the purpose of concentrating the silver in the lead, is founded on the fact that lead, if alloyed with silver to a certain proportion, is more liquid than pure lead, or lead alloyed in a very small proportion with silver. The advantage of this process is the reduction of the quantity of lead, which otherwise would have to be separated from silver by cupellation.

When a sufficient quantity of lead, containing silver, is melted in a cast-iron vessel and uniformly cooled, small crystals will form, increasing in quantity. These crystals, when taken out by means of perforated ladles, will be found a great deal poorer in silver than the fluid remainder in the kettle, which again will be found richer than the original lead.

On account of the adherence of fluid lead to the crystallized, the separation cannot be perfect, but repeated recrystallization of the crystals results finally in two sorts of very poor and rich lead. There is, however, a limit, beyond which no concentration to a higher degree can be effected.

According to Professor Reich, the melting temperature of lead, containing 0·0065 per cent. of silver is 610° Fahrenheit. Lead, containing 0·476 per cent. of silver melts at 588°; but lead containing 2·25 per cent. of silver melts at the same temperature as pure lead, consequently no crystallization of poor lead can take place. Alloys of equal parts of both metals, or three parts of lead to one of silver, require a higher temperature of melting than pure lead.

The following table shows the proportionate progress of the enrichment of lead, by crystallization of the poor:

Amount of silver in pig lead.	Amount of silver dipped in crystals.	Amount of silver in the liquid remainder.
0·704 per cent.....	0·390 to 0·466........	1·025
0·732 "	0·318 " 0·374........	1·076
0·966 "	0·410 " 0·680........	1·450
0·988 "	0·390 " 0·624........	1·530
1·442 "	0·682................	1·922
2·090 "	2·011................	2·260
2·206 "	2·216................	2·246
2·206 "	2·212................	2·264

The good result of this manipulation depends—

1. On the right management of the temperature. If the temperature be too low, no separation can be obtained, and if on the other hand the lead is too hot no perfect formation of crystals can take place.

2. On the right quantity of lead which is taken into operation. It requires at least two tons and a half, in order to effect a slow change from the fluid into solid condition, affording thus sufficient time to remove the crystallized lead.

3. On the number and size of iron kettles, depending on the quantity of lead designed for this process and on the amount of silver in it.

The right temperature, the time of crystallization, and the preservation of temperature must be found by experiments. The use and advantage of this method depends principally on the quality of the lead and the quantity of silver in it. The process will succeed if the lead is free from other base metals. Antimony and copper aggravate the formation of crystals, from which the liquid lead separates with difficulty. The production of pure lead, also an object of this method, cannot be obtained in that case. If, therefore, such impure lead is designed for crystallization, it must be subjected first to purification, causing thus expense and loss in metals. This purification of impure lead, especially if antimony is present, may be executed by melting it in a reverberatory furnace at a very low temperature, keeping it for some time in such condition. It forms then a crust on the surface, which principally consists of oxyd of lead and antimonate of oxyd of lead. This crust must be drawn off as long as it appears, or stirred by increased heat with addition of lime and charcoal dust, by which a great part of the lead is reduced again. But it is evident that under such circumstances, even when the purest lead would result from subsequent crystallization, a direct cupellation in California and Nevada Territory is more advantageous, and that, if the cupellation on account of too small amount of silver would not pay, the crystallization, depending on the purification of lead, would pay still less.

If the lead is not overloaded with base metals, but containing, however, so much as to interfere with the crystallization, a more simple refining may be adopted. The lead is introduced into a Pattinson kettle melted and stirred with a wooden rod, by which the lead is brought into a turbulent motion, exposing always a renewed surface to the air, promoting thus the oxydation of the base metals. The impurities are drawn off, till they cease to appear.

The number of kettles required for crystallization is determined principally by the amount of silver. The process offers the best advantage, if the lead contains from five to ten ounces per ton. In this case, a few crystallizations render two kinds of lead, one poor, the other rich; the latter ready for cupellation. At a higher amount of silver the process is prolonged, causing more expense for fuel and labor; but under favorable circumstances, for instance, if very pure lead is obtained which would command a better price, then also lead containing from fifty to sixty ounces per ton could be advantageously subjected to the concentration process.

The degree to which the concentration is carried on is important, because, besides the increased expense for labor and fuel, a very rich lead produces a richer litharge when cupelled. The concentration accordingly ought not to exceed four or five hundred ounces per ton.

Lead, containing silver in such small proportion as will not pay the expense of cupellation, can be worked advantageously by way of crystallization. The low temperature does not much affect the lead, and besides when the lead is pure the loss in silver is insignificant. The loss of lead in England, which is generally pure, amounts to two per cent. Other qualities of lead, in Germany, suffer a loss of three per cent., while the loss in cupellation and the reduction of litharge amounts to from eight to ten per cent. Generally there is no gain on fuel or labor in this process, compared with cupellation. If there is a considerable amount of silver in the lead, the crystallization may be even more expensive, but the gain in lead may cover the difference and leave also a profit. This, however, depends on the price of lead.

CHAPTER VI.

DESCRIPTION OF MACHINERY AND FURNACES.

COMMON IRON PAN.

Sec. 49. Fig. 22 represents a common iron six-foot pan. Fig. 23 is a vertical section of Fig. 22 on the line AB; a shows a wooden cross, to which the wooden block, b, with the iron shoes, c, are fastened by the bolts, d. Each shoe has a pin, e, about one inch long, fitting in the wooden block, in order to prevent movement.

On the shaft, g, is the yoke, f, fastened by a key. The two ends of the yoke fit in the holes, h, of the cross, a, but not too tight, so that the muller can follow the wear of the shoes. Pans, having the gear underneath, and the shaft through the cone, i, are so arranged that by means of a screw the muller can be raised. This arrangement for raising the muller is not important, as the muller generally grinds with its full weight. The steam is introduced into the pulp, through the pipe, l; $k, k,$ are the discharge pipes; m represents the false bottom, made of one piece, two inches thick. This bottom must be one inch less in diameter, so that half an inch of free space is left between the bottom and the pan on the sides, and on the cones. The best way to fasten these bottoms, and also to prevent the quicksilver from getting under them, is the following: Strips of strong cloth, two inches and a half wide, are laid over the space between the bottom and sides, which is filled with a paste of iron filings and wedged with well dried wooden wedges, quite close together, so that the cloth is equal in height on both sides of the wedges, which are driven in tight. The wedges must be a little shorter than the thickness of the false bottom, leaving thus a space above them, which is covered with a paste of iron filings.

WHEELER'S PAN.

SEC. 50. Fig. 24 shows the ground view of the pan, and Fig. 25 the vertical section of Fig. 24, on the line $A B$. For the sake of clearness of representation, the yoke, a, and the guide-blade arrangement, q, b', b, c, of Fig. 25, are not represented in Fig. 24, but the position of the guides is shown by dotted lines, c. The muller, d, and the ring, e, (the two journals of which move in the box, f, fixed to the muller, while the other two, g, move in the box, h, of the yoke, a) cover only half of the pan, Fig. 24. The other half shows the dies, i, laid directly on the bottom of the pan, k. They are kept in place in the centre by the ring, j, and on the sides by the inclined ledges, l. Each die has for this purpose a projection, m, which is placed under the ledge, l, with the beveled side, as represented in Fig. 25, m'. The dies are one inch thick, beveled on each long side in the same direction, so that, in putting them together, the groves, i', are formed.

In Fig. 25 the muller, d, shoe, n, and die, i, are represented on one side. The muller has twelve oblong openings, d', two and one-fourth inches wide by ten and three-fourths long. One of the long sides towards the dies is beveled, as shown by the dotted lines, s. The projection on the shoes is of the same shape, being only half an inch narrower and as much shorter, so that the space of half an inch is wedged with dry pine wood, n', by which the shoe is fastened to the muller. The shoe below the muller is represented by the dotted lines o. On the outer ledge of the muller are inclined ledges, o', which, in connection with those of the pan, l, create an upward current of the pulp; c, c, are guide-blades conveying the pulp to the centre. These guides have at the outer end a projection, like a hook, as indicated by the dotted lines c, c. This hook catches the blade, p, which is also bent hook-like, fastened to the pan by an iron wedge, between the ribs, p'. Thus the guides resist the current of the pulp.

The guide-blades, c, are connected with the frame, q, formed of four rods, screwed to the ring, b', which again is in connection with the lower ring, b, by four bolts, to which the guides are attached. The **frame** rests by means of the screw, q', on the shaft, r, and can be raised or lowered.

a is the steampipe, conveying the steam directly into the pulp; u shows the steam chamber, if the pulp is to be heated through the bottom. This, however, has proved unnecessary, and it consumes more fuel. On that account no more pans of this kind are made with a steam chamber. As a consequence, the arrangement with the box, t, is also altered, and affords more convenience in oiling and handling; w and w' show the apparatus for raising the muller. By screwing the rod, y, the muller will be raised gradually, but if a sudden lift is required, the rod must be pushed down.

The dies, as well as the bottom of the pan, are not so perfect as to fit exactly without giving way a little in one place or another, thus presenting an uneven grinding surface, and causing a jarring for many hours after starting. To avoid this, some wet, muddy tailings, or ore, is introduced into the pan and spread uniformly, nearly half an inch deep. Each die is then laid in the proper radial direction, the projection of the outer end under the ledge, l, and imbedded well into the mud, till it lies solid and level. Care must be taken that the stump edge 1, Fig. 24, does not project over the sharp edge 2 of the next die. When all twelve dies are set in, the collar, j, is put over the heads of the dies, and fastened by turning it under the nuts of the centre-piece 3. The spaces, i' and k, are filled with wet ore, level with the surface of the dies, and the muller placed over them.

Some water is poured over the muller, then some diluted mud, and allowed to rest for about two hours before starting, in order to effect the swelling of the wedges, n'.

Above the pans a sliding block is required, for the purpose of lifting the muller. The muller must be lifted at least once every week, and pan and muller cleaned from adhering amalgam, which accumulates round the centrepiece, preventing the ore from passing freely under the muller. The first three or four charges require two or three hours' longer grinding, on account of the roughness of the shoes and dies. These pans are made in the Miners' Foundry, San Francisco.

WHEELER'S AGITATOR.

Sec. 51. Fig. 26 represents a vertical section of the agitator (exclusive of the arms, a, which are shown in a front view). The upright shaft, b, is hollow for the purpose of conveying the water through the arms a, of which there are four, into the pulp, in order to dilute it for an easier separation of the amalgam from the sand. If, however, it is treated as described in Sec. 27, no dilution is required, therefore the hollow shaft and arms are superfluous and may be replaced by solid ones. In case the pulp is too thick, some water may be added in the pan, several minutes before the discharge.

The shaft, *b*, slides in the gear, *c*, on the fixed keys, resting on the fork, *d*, and a loose collar, *e*. By means of the lever, *d*, and screw, *f*, the shaft can be lifted with the shoes gradually. The cast-iron bottom, *g*, is inclined towards the centre, ending in a bowl, *h*, for the reception of quicksilver. This bowl is kept always full of mercury so that when the pans are discharged the superfluous quicksilver runs out by the siphon, *i*. The three-eighth pipe, *k*, four inches above the bottom, carries out the tailings constantly, being diluted with a one-inch stream, *m*, of clear water. The lower, one-inch pipe, *l*, serves for the discharge, if cleaning is required; *o* is also a discharge hole; *p* represents the shoes; they are light and fastened with bolts so that little friction takes place; *q* are wooden staves from twenty-five to thirty inches high.

Hepburn's agitator differs from Wheeler's in having a level bottom; and instead of the bowl in the centre, there is a circular groove near the middle of the iron bottom, connected with the siphon. This agitator will be worked to the best advantage without dilution in the same way as Wheeler's (Sec. 27).

HEPBURN AND PETERSON'S PAN.

Sec. 52. Fig. 27 represents a perpendicular section of the ground view of Fig. 28, on the line *A B*. Fig. 28 shows a ground view of the pan, the centre piece of which is a horizontal section on the line *C D* of Fig. 27.

The pan, a', is eighteen inches deep on the side, and twenty-seven inches deep near the cone, b, forming thus an inclined bottom. The shaft, c, goes through the hollow cone, b. The cylindrical driver, d, is fixed to the shaft by a stationary key. The muller, m, has three stands, e', joining in one piece, e'', in which three openings, p, are made, for the passage of the logs, e, of the driver, d. When the driver is set in, it must be turned to the right, so that the logs, e, will come under the leveling screw, x, as represented in Fig. 28 by the dotted lines, joining the logs, a, of the stand, e'', whereby the motion is transferred from the driver to the muller. The screw, f, serves for raising the muller. The nut, o, is movable inside of the driver, but is fixed in the handwheel, h, so that the motion of the screw, f, can be arrested by holding the handwheel, h. Holding this handwheel, the muller can be raised when in motion, by the screw, f, with the upper hand-wheel, h'. The upper part of the cone, which serves as a box for the shaft, c, is lined with babet metal, z. The shoes, n, are fastened to the muller by the bolt, m'; there are also two pins on each shoe, which fit in small holes of the muller, to prevent motion. In Fig. 28, these shoes are shown by dotted lines. The dies, i, have two projections on the lower side. The projections fit into corresponding grooves of the pan bottom. A wedge, r, fastens the die in its place. The pipe, q, serves for the discharge, and q' for conveying steam.

These pans are furnished by the **Vulcan Foundry,** San Francisco.

ROASTING FURNACE.

Sec. 53. The roasting furnace, represented by Fig. 29, the horizontal section of Fig. 30 on the line CD; then Fig. 30, a vertical section of Fig. 29 on the line AB shows the right proportion and dimensions of an economical furnace in regard to fuel and time of roasting. The hearth-bottom, a, is made carefully of the hardest bricks, laid edgewise as close as possible. Bricks with uneven sides must be rejected; b shows a square hole in the floor for discharging the roasted ore. The flue, c, communicates with the holes, c', in the arch, v, nine inches in diameter; they being perpendicular, are not liable to be choked by the ore; d is a canal for the purpose of cleaning c. The distance between the arch and bottom near the bridge, g, is twenty-one inches, but near the flue, c', only eight inches. The flue, c, can be led, either under the floor, or directly, or through dust chambers, into a chimney, which, for one furnace, must be twenty-five or thirty feet high, with the inside from sixteen to eighteen inches square. On top of the chimney an iron cover, controlled by a chain, regulates the draft of the furnace; e represents an iron funnel, big enough to receive one charge of ore. It is fitted tight to another small cast-iron funnel with an opening of about four inches, with a slide, i. The canals, f, are draft holes for the purpose of drying the brick work.

As the temperature does not amount to white heat,

common bricks are generally used for roasting **furnaces**. There are often, however, such bricks which **cannot** stand **even** the roasting heat, especially in the fireplace. **New** bricks, therefore, must be examined first by exposing one to light red heat for an hour or so.

The bridge, g, is the part most exposed to injury by fire on one side, and by stirring with hoes on the other. It is of great advantage to make the bridge of one or two pieces of some kind of sandstone, granite or conglomerate, which does not burst on being heated.

The furnace must **be well** tied with iron rods, k, and uprights, l. In some places, the uprights are made of wood, six-by-six or six-by-eight. A new furnace must be dried carefully. It requires five or six days' slow fire. Upon this first drying the durability of the arch depends. The furnace must always be red-hot before the first charge **of ore is introduced**.

MECHANICAL ROASTING FURNACE.

Sec. 54. Fig. 31 represents a vertical section of a mechanical furnace. The iron hearth-frame, a, twelve feet in diameter, has sides ten inches **high**. The fire tiles, b, are first laid, then the **four-inch** bottom, c, formed by two rows of bricks; d is **the** discharge opening, three feet three inches long, four inches wide, with **iron** door on hinges below. The funnel, e, conveys the discharged ore through the hole, f, outside **of the** furnace. The hearth is set in motion **by the cog-wheel**,

g. Five or six revolutions will be sufficient. The hearth revolves on eight rollers, h, and on the balls, h'. The pin, i, prevents any motion sideways. The fire tiles, k, on the bridge are placed close to the hearth. The cast-iron pipe, l, rests outside, independent of the furnace, on both sides. It lies above the centre of the hearth. This pipe (three and one-half by two inches inside) cooled by a constant stream of water, withstands destruction by heat, cooling also the iron stirrers, n. There are eight or ten of these stirrers on the pipe, so arranged that while one-half of them turn the ore to the right, the other half in a reversed position throw it to the left. The doors, m (on each side), must be wide enough to allow the taking out and replacing of the stirrers. This can also be effected by the other door, o. The flue, p, is indicated by dotted lines.

The ore is introduced as usual by the funnel, p', while the furnace is red-hot and the hearth revolving. After roasting, the hearth is stopped, the door of the discharge hole, d, opened, and the hearth started again. When the ore is discharged, the door must be closed and the space, d, filled with roasted ore.

THE RETORT.

SEC. 55. The retort, Fig. 20, represents a front view without the door. Fig. 21 is a vertical cut of Fig. 20 on the line AB, being shut by the door or cover, a. The retort is made of cast-iron, four feet long, eleven inches wide and nine inches high. The retort has two wings, b, as long as the retort, on which it rests on the mason work, so that three inches on each side, d, are left clear. The flame, on account of the wings, will play round the lower part of the retort, passing the space, c, then f above the retort, escaping through g. The fire-place, h, has grates two feet long, and may be according to the fuel either three feet long or shorter, as indicated by the dotted line, i. The condensing pipe, k, is furnished at the end with a funnel, l. A sheet-iron pipe, m, rests on the funnel, made tight by some clay or mud inside. Through the lead pipe, n, a constant stream of cold water flows to the funnel, then round the pipe, k, up and out at o. The funnel is wrapped with cloth, p, which reaches into the water. The water level, q, must be kept about half an inch below the funnel.

THE CRUCIBLE FURNACE.

SEC. 56. This furnace is very simple. Fig. 13 is a vertical section of Fig. 14 on the line $A\,B$. Fig. 14 represents a ground view. This furnace, calculated for the black-lead crucible No. 40, is fifteen inches square, lined with fire brick; a shows the slide doors, c the cast-iron plate with the rib, d, to support the doors, a. The crucible, e, stands on a piece of fire brick on the grate, f, which consists of movable iron rods; g is the ash-pit, having a very tight floor, or a cast-iron plate with a rim on three sides.

A round or circular furnace for the same size of crucible must be eighteen inches diameter in the clear. As a matter of course this furnace will serve also for smaller sizes of crucibles.

THE MELTING FURNACE.

SEC. 57. Fig. 33 represents a horizontal section on the line AB of Fig. 32, which is a vertical cut on CD of Fig. 33. Another vertical cut on the line EF is shown in Fig. 34. The cast-iron pan, a, is thirty-seven inches in diameter and fifteen inches deep. This pan serves for the reception of the fire-proof material, e, forming the crucible. This pan has a tap-hole, b, close to the bottom about six inches in diameter. On one side near the rim, the pan has a shoulder, c, on which the flue-plate, d, rests.

The pan, a, is placed on bricks, g, about four inches apart so that air canals, h, are formed, for the purpose of cooling the hearth-mass, e, in the pan. The sides of the pan are also exposed to the cooling air. The flue-plate, d, has a pitch of about thirty degrees. The hearth, e, is beaten to six inches above the edge of the pan, in which a crucible twelve inches deep is cut out with a curved blade, so that under the crucible in the centre, the mass is about six inches thick; i represents the feed door, through which the ore is introduced on the inclined flue, l, whence the melting mass runs into the crucible, f. The slag flows through the canal, m, three inches wide and nearly four inches deep.

The tap-hole, *b*, is nine inches on the outside, ending in the centre of the crucible with a half-inch hole, *n*. The door, *o*, on the slag canal, *m*, serves for drawing off the slag; *p* shows the dust chamber; *q* and *q'* openings for the purpose of cleaning the chamber. The ore particles, striking against the partition, *r*, drop down, forming masses of slag-like stalactites, which must be removed from time to time.

The arch, *x*, formed of fire bricks (like all mason-work coming in contact with fire), must be taken off when a new crucible is required. The crucible may last two or three weeks according to the material.

When all is finished, except the roofing, the flue-plate, *d*, is lined first with one row of fire brick, *y*, after which the hearth-mass is beaten in the pan in the following way: the mass is prepared of fire-proof material, either of fire-brick previously used, carefully cleaned from slag, pulverized, and three parts of it (in volume) mixed with one part of good clay, or three parts of pure white quartz with one part clay; or three parts burnt fire-proof clay with one part unburnt will answer the purpose. The materials have to be crushed dry and sifted through a sieve of sixty-four holes to the square inch. After being well mixed, the stuff is sprinkled over with water while it is shoveled on a clean floor. The shoveling must be performed with much diligence, from one side to the other in order to have the whole mass uniformly moist.

The quantity of water must be found out by the

appearance of the mass. A handful of it, squeezed hard, must form a ball, which may be handled gently, but it should crumble to its former state under a pressure of the fingers. With this prepared mass the pan, a, is charged about six inches deep, leveled, and by one or two men beaten in, by means of iron bars of about twenty pounds weight and five feet long, having at one end exactly the size and shape of a hen's egg at the larger end. The men commence to beat the hearth in the centre, pursuing a spiral course, stamping with the egg-point in a perpendicular line, by a lift of about eight inches, striking with the rod close to each preceding stroke. When, by a screw-like advance, the stamping has reached the side of the pan, it can be carried on back to the centre in the same way, then again to the side and so on, till about two inches loose mass remains.

If a hole can be scratched easily with the finger into the stamped mass, the rod must be used with more force. Another charge of the same quantity of stuff is introduced as before. Care must be taken to have always two inches of loose mass above the stamped, before the next charge is put in. If the whole mass be beaten hard, the next charge cannot unite with the underlayer perfectly. The beating on the second charge is executed in the same way as before and so on, till the hard-beaten mass stands about two inches above the line v, extending into the flue as far as shown by the drawing. The hearth-mass in the flue, l, must be stamped with lighter wooden stamps of the shape of the iron ones.

The hearth is then cut level, as indicated by the line, v, and the surface, l, in the flue. After this, the crucible, f, must be first cut roughly with a hatchet or a similar instrument, but the finishing to the proper size and shape is done with a curved blade. The tap-hole, b, is cut out as represented in Fig. 34.

When this is done, the arch and binding of the furnace will finish the work. It is now important to dry the hearth carefully. It is very advantageous to cover the surface of the crucible about three inches thick with ashes in order to prevent the direct action of the fire on the mass. After three or four days' drying, the fire may be increased by degrees, so that the furnace becomes red-hot after five or six days' drying.

The chimney should be forty or fifty feet high, from eighteen to twenty inches square in the clear, and lined with fire-bricks.

The fire-place, H, is calculated for coal

CUPELLING FURNACE.

SEC. 58. The cupelling furnace is represented by Fig. 35, which is the vertical section of Fig. 36 on the line $A\,B$. Fig. 36 shows the horizontal cut of Fig. 35 on $C\,D$. The brick-work, b, is one foot eight inches above the floor. On this block are laid rows of bricks, c, forming channels, d, for the free circulation of air. On the brick bridges a cast-iron pan is placed for the reception of the hearth-mass.

This pan is four feet in diameter with a throat, f, ten inches long and twelve inches wide. The bottom is perforated with several holes, each half an inch in the clear, for the easy escape of moisture. This pan is eleven inches deep; g is a cast-iron plate on which the litharge falls; i represents a fire-tile on which the lead bars are placed; it is a little inclined towards the test, k. The opening, l, is for the nozzle of a pair of bellows. The nozzle, a, Fig. 37, is also inclined at such a degree that the stream of air may touch as much of the surface of the bath as possible. A continual stream of air produced by a fan answers the purpose better than a forge-bellows; n is the opening above the litharge-bridge, f. Through this opening the lead vapors and all burning products escape.

The arch, o, is made of common bricks and must be removed as often as a new test is required. Larger furnaces are provided with a movable arch, formed of sheet-iron or flat wrought-iron, lined with clay inside. By means of a crane the arch can be raised and turned to one side.

If the mass, e, forming the hearth, is composed of wood ashes, it causes much trouble in cupelling. Bone-ash is an excellent material, but too expensive. The most proper material is marl. If this cannot be obtained, a composition of pulverized lime-rock and clay will answer perfectly. For this purpose one part of clay, or in the absence of it, light yellow loam is mixed with three parts of lime-rock, all well dried and pulverized.

This mass must be sifted through a sieve of about one hundred holes to the square inch, then sprinkled with water and prepared as described in Sec. 57.

The pan, *a*, is then charged about six inches deep with the prepared hearth-mass, after the front opening has been shut up securely in some way, for instance, by bracing, as indicated by *x*. One or two men commence to stamp the mass in the centre with wooden or iron stamps of the same size and shape as described in Sec. 57.

The stamping goes on, till the hearth is beaten about three inches below the rim, when a cast-iron ring of the size and shape of the pan, but with sides only six inches high is placed rim to rim, corresponding on the pan, as indicated by the dotted line, *z*, Fig. 35. Some more mass is then introduced, and the stamping continued till the hearth is formed hard two or three inches above the rim of the pan, *a*.

The mass is then cut with a knife around the ring down to the pan's edge, in order to prevent the breaking of the mass when the ring is removed. The hearth is cut level with the line, *p*, then the test, *k*, five inches distant from the rim, with a curved knife. This is six inches deep and four feet diameter. The litharge-bridge, *f*, should be cut inclined, as represented by *r*.

When this is finished, the arch can be formed and the furnace slowly dried.

The cast-iron pan, *a*, offers great advantages in the preservation of the test, in less danger in drying and in withstanding the heat longer, which has less influence on the mass, being cooled from beneath.

THE REFINING FURNACE.

SEC. 59. For the purpose of refining silver or melting retorted amalgam, the furnace, Figs. 32, 33, will answer perfectly, changing a few dimensions. The fire-place, H, if intended for dry wood, should be fifteen inches wide and three feet six inches long. The crucible, f, should be only six or eight inches deep, the line, v, being then one or two inches above the rim of the pan, a, as represented by the dotted line, f'. The flue, l, will answer with a length of from twelve to fifteen inches, with a door, i', for the purpose of introducing the silver on the inclined plane, from which it melts into the crucible. For the hearth-mass, if a purer sort of silver is treated, the same mixture as described for this furnace (Sec. 57) may be used; but in treating silver which contains lead, or which requires an addition of lead for the purpose of oxydizing the base metals or other impurities, the mixture of lime and clay as used for the cupelling furnace (Sec. 58) is preferable.

There is no flue-plate, d, required, also no tap-hole, as it is more convenient to dip the silver out with an iron ladle into moulds.

PART SECOND.

GENERAL METALLURGY OF SILVER ORES.

CHAPTER I.

DIVISION AND ASSAY OF SILVER ORES.

SEC. 60. Silver ores extensively used for extraction are generally:

1. Those in which the silver appears as one of the principal metals. To this class especially, belong:

Native silver, containing sometimes as much as three per cent. of antimony, arsenic or iron.

Antimonial silver with seventy-seven or eighty-four per cent. of silver.

Tellurid of silver with sixty-one per cent. of silver, containing sometimes gold and traces of iron.

Silver glance with eighty-seven of silver.

Brittle silver ore with seventy, light-red silver ore with sixty-five, and dark-red silver ore (rubysilver) with fifty-nine per cent. of silver.

Silver fahlerz with from eighteen to thirty-one per cent. of silver and from twenty-six to fifteen per cent. of copper; light-gray silver ore containing 5·7 per cent. of silver, thirty-eight of lead and traces of copper. The gray silver ore from Freiberg (Saxony) has from 29·73 to 32·69 per cent. of silver.

Stromeyerite (silver-copper glance), fifty-three silver and thirty-one copper.

Polybasite with from sixty-four to seventy-two silver, and ten to three copper.

2. Such ores, in which the small and variable amount of silver appears with other useful or not useful metals.

Sulphuretted minerals are always rich in silver, than the oxydized. The ferriferous sulphides are the poorest, then follow sulphides of zinc, of lead and then those of copper.

To this class belong:

a. Argentiferous Lead Ores.

Sulphuret of lead or galena containing generally from 0·03 to 0·01 per cent. of silver; often also 0·5, but seldom one per cent. In some cases, however, galena is found to contain as much as seven per cent. of silver. The amount of silver in this ore varies much even in the same mine. Fine-grained galena is generally richer than coarse-grained or crystallized, although in Arizona the latter proves to be richer than the fine-grained.

Carbonate of lead is generally found poor in silver.

b. Argentiferous Copper Ores.

Fahlerz (gray copper) contains traces of silver up to thirty-one per cent.

Copper pyrites, purple copper and copper glance are poor in silver. In such cases the silver is extracted only when this ore occurs in company with lead ore, but in some cases the sulphide of copper and the sulphuret of silver form much richer combinations than the sulphuret of lead.

c. ARGENTIFEROUS ZINC ORES.

The zincblende (sulphuret of zinc) contains traces of silver up to 0·88 per cent. The carbonate of zinc shows only traces.

d. ARGENTIFEROUS IRON AND MAGNETIC PYRITES,

Antimonial and arsenical ores, then bismuth, cobalt and nickel ores are poor in silver.

The ores of the first class are seldom found in large quantities or massive. They are generally disseminated in coarser or finer particles through the gangue, and in that case circumstances of the locality will decide whether it is more advantageous to concentrate the ore, or to save the loss of silver suffered by concentration, and to beneficiate a larger bulk with more expense.

The ores of the second class require particular attention, partly as to a close separation of the different metals in regard to concentration, and partly in relation to a suitable mode of extraction for each kind of ore.

SEC. 61. To find out the amount of silver in the ore,

there are two different tests in use: **the** dry or fire assay, and **the wet** assay. In the fire assay **the** silver of the ore is fast taken up by the lead and then separated by cupellation. In consequence of volatilization and oxydation of the silver, the **dry** assay suffers a certain loss. The oxyd **of silver with that of** lead draws into the cupel **mass, and loss increases with** the larger amount of silver; **but if the lead is not rich,** the loss can hardly be determined by the balance. The loss by cupellation will also increase if too much heat is applied. Assaying poorer ore, and using proper heat, **the** loss of silver is generally not taken into account; but if rich alloys are assayed, the loss of cupellation must be added to the weight of the button, according to tables containing such losses, the results of careful experiments.

Small amounts of silver are easier and more surely found by the **dry than by** the wet assay, but the latter offers more exactness in rich alloys. For this reason the wet assay is used in the mints and such offices, where silver bars and rich alloys have to be assayed.

A. DRY SILVER ASSAYS.

SEC. 62. As mentioned before, **the dry assay is based on the** combination of the **silver in the ore, with** the lead, **which was** added either in metallic condition, or as litharge **or acetate of** lead. All lead or litharge of commerce **contains some silver, but** generally so little that its influence on the **assay is** insignificant. For important assays of substances which contain but very little silver, the lead must be prepared from acetate of lead, or the latter is used directly after calcination. The calcined mass contains more carbon in the acetic acid than required for the reduction of the oxyd of lead.

In most instances the test contains foreign matters, which, to be dissolved, require fluxes. As fluxes, potash, soda, borax, salt (Glauber's) and saltpetre may be used. For the reduction of oxyds, black flux, flour or charcoal dust will serve. During the cupellation of argentiferous lead, the lead is first oxydized, transferring oxygen to other base metals, the oxyds of which together with the litharge, are taken up by the porous cupel.

The following assays may be distinguished:

I. PROCEEDING FOR ORES, MELTING PRODUCTS, ETC., WHICH ARE NOT ALLOYS.

1. ASSAYS FOR RICH SILVER ORES.

a. Melting with Metallic Lead (Scorification) and Cupellation of Lead.—One hundred parts of the sample, containing sulphur, antimony, arsenic, and earths are mixed with four hundred parts of test lead in a scorifier (clay vessel) and covered with four hundred parts more of lead. This proportion of lead and ore, eight to one, is generally sufficient, but if there is a considerable quantity of zincblend in the ore, twelve to fourteen, and in presence of much copper or tin, twenty to thirty times as much lead is required, as the weight of the ore. An addition 0·1 or 0·16 parts of borax is necessary then, when the ore contains a great deal of earthy matter or zincblend. The charged scorifier is introduced into the red-hot muffle, the muffle closed and strong heat applied for fifteen or twenty minutes. Then the muffle is opened and the melting continued till the lead becomes entirely covered with lead slag. The muffle is then closed again for fifteen minutes, then the scorifier taken out, and its contents immediately poured into a circular ingot mould. When cold, the lead button is hammered square, the sharp edges and corners hammered down and the button finally cupelled.

The lead button is placed on the red-hot cupel and the heat raised. As soon as the lead is perceived to work, the muffle is opened again to admit the air. The cupellation must be executed at the lowest possible heat. The guides for the right temperature are the rising lead fumes and the glowing of the cupel. If the fumes rise slowly to the middle of the muffle and the cupel glows reddish-brown, it indicates the right temperature; but if the fumes disappear immediately above the cupel, which appears light red, the temperature is too high. In this case, some broken pieces of crucible may be introduced in the rear of the cupels. The temperature must be raised again at the end of the operation. The cupel is then removed and the silver button weighed.

b. Melting of Argentiferous Lead Ores and Melting Products without addition of Lead, and Cupellation of the Regulus.—Ores or other substances containing sufficient lead, and not rich in silver, are melted like lead assays. The obtained lead button is weighed and then cupelled. This method is applied to ores containing not more than about fifty ounces of silver per ton and not less than six hundred pounds of lead, also lead-matt, hearth, scrapings, litharge, etc.

c. Melting of Unroasted Ores with Litharge or Acetate of Lead and Fluxes.—To these assays, ores are subjected, not too rich in silver, containing sulphur, antimony, and arsenic. The oxygen of the litharge oxydizes the sulphurets, while the disengaged lead combines with the silver. Another part of the litharge serves as a flux, and dissolves the earthy matters. To obtain a better fluxing, other fluxes must be added, as: borax, glass, soda, or potash. The addition of coal dust, flour, or black flux reduces more lead and favors the desilverization of the ore. One part of ore is mixed with from four to eight parts of litharge, two parts potash or soda and some calcined borax, also 0·06 charcoal dust. This mixture is placed in a crucible, covered with salt and melted. The heat must be kept moderate, till the mixture is melted down. The temperature is then raised nearly to white heat for about fifteen minutes, whereupon the crucible is taken out, and when cold broken. This method is not quite so perfect as the scorification, *a*.

d. Melting of Roasted Silver Ore with Litharge and Black Flux.—This method yields always less silver than the preceding assays, on account of the loss of silver in roasting.

2. ASSAYS FOR POOR SILVER ORES.

Generally, the assays for rich, are adapted also for the poor ores. The procedure differs only in this that a little larger quantity is weighed out, and that several assays are made of one sample, the lead buttons of which are cupelled into one silver grain.

a. Ores rich in Sulphurets, Arsenical or Antimonial Combinations.—For instance, iron pyrites, arsenical pyrites, blends, etc., are best melted with litharge in crucibles 1 c. The scorification is less suitable, unless the test lead is quite free from silver. Very poor galena may be melted with thirty or forty per cent. of saltpetre and one hundred per cent. of soda, or better one hundred per cent. of litharge. Also one ounce galena with one and a half ounces black flux, 0·1 ounce iron filings, and five ounces of soda, or with 0·5 ounces saltpetre, 7·5 litharge, and five ounces of soda.

b. Ores containing a great quantity of Earths.—For instance, very poor tailings, slag, etc., are assayed in the following manner: Half ounces of the sample are weighed out from six to eight times and each separately mixed with 1·5 or two ounces of soda and 0·05 charcoal dust, then introduced in so many crucibles, each covered with half an ounce granulated test lead, and finally with salt. The crucibles are then exposed to one or one and a half hours' moderate melting, after which a strong heat is applied for fifteen minutes more, and the assays taken out. The six or eight lead buttons obtained by this melting are laid in scorifiers under the muffle, and by oxydizing concentrated into one button, which is then cupelled. At the same time, six or eight ounces of the same test lead must be concentrated and cupelled and the amount of silver obtained subtracted from the assay. It may be necessary to concentrate ten or more assays into one when the ore is very poor. An assay of slag can be melted with nine hundred per cent. of litharge and 0·8 per cent. of quartz without other fluxes.

II. PROCEEDING FOR ARGENTIFEROUS ALLOYS.

Argentiferous lead, pig lead, and bismuth are subjected directly to cupellation in quantities of one-half or one ounce. If poor in silver, several ounces may be concentrated by scorification before being cupelled.

Argentiferous tin and zinc must be first oxydized under the muffle, then scorified with sixteen parts of test lead and four parts of borax, or they may be melted in a crucible with litharge and black flux.

Argentiferous wrought-iron, cast-iron, or steel are first oxydized by means of nitric acid, then the dried oxyd scorified with eight to twelve parts lead, two or three of borax and one of powdered glass. The lead is subjected to cupellation.

Argentiferous copper, black copper, brass, etc., are first broken in small pieces, then one-fourth of an ounce of either scorified with sixteen to thirty parts of test lead, and then cupelled.

Cupriferous Silver.—Before a proper cupellation can be carried out, it is necessary to know the aproximative alloy of copper. For this purpose the test is first cupelled with ten or fifteen parts of lead, and the quantity of copper ascertained. Pure silver should be cupelled with 0·3 of its weight of lead; an alloy of fifteen silver and one copper requires three times as much lead; fourteen silver and two copper should be cupelled with seven fold; twelve silver and four copper with ten fold; eleven silver and five copper with twelve fold; nine silver and seven copper with fourteen fold, and eight to one silver and eight to fifteen copper with sixteen or seventeen fold quantity of lead.

The cupel absorbs about as much litharge as its own weight, whereby the size of the cupel is indicated for a certain quantity of lead. The copperous silver is wrapped in a piece of paper or lead-foil and introduced into the glowing cupel, then the lead in one piece, or the lead may be first brought into the cupel, and then, when it becomes bright, the alloy added. As soon as the test commences to work at a strong heat, the temperature must be diminished, and only at the end of the operation increased. The regulation of the temperature is effected by closing or opening the draft holes.

B. WET ASSAYS.

Sec. 63. This method (Gay–Lussac's) is very suitable for rich alloys and gives the most accurate results. A determined weight of silver is dissolved in nitric acid, and precipitated by a graduated solution of common salt. The calculation is then made on the quantity of the solution used for the precipitation. One hundred cubic centimetres of the standard salt solution precipitate one gramme of silver.

C. BLOWPIPE ASSAYS.

1. Argentiferous Ores and Substances which are not Alloys.

Sec. 64. The fine pulverized test is mixed with fine-sifted lead (free from silver) and with borax glass, then melted, and the lead cupelled. This assay consists of the following operations:

a. The Dressing of the Assay.—The required quantity of borax-glass depends on the fusibility of the substance. For assays that fuse with difficulty, for instance, iron or arsenical-pyrites, nickel or cobalt ores, one grain of borax is sufficient for one grain of ore. The quantity of lead depends on the quantity of substances which are difficult to scorify, for instance copper, but especially nickel and cobalt. If there is not over seven per cent.

of copper or ten per cent. of nickel, five grains of lead to one grain of ore will answer. Ten or twelve per cent. of copper (bournonit, eucairit, tin pyrites) require seven grains of lead. Thirty to fifty per cent. of copper (silver-copper glance, copper pyrites, copper blend, fahlerz, tenantit or copper matt) must be mixed with ten grains of lead. Sixty-five–seventy per cent. of copper (purple copper) require twelve grains and eighty per cent. of copper (copper glance) fifteen grains of test lead.

Silver sulphurets, roasted lead, and silver ores, amalgamation and extraction tailings, bromic, iodic, chlorobromic silver, hearth, argentiferous slag, etc., are dressed with five grains of lead 0·5 grains of borax-glass and the obtained lead button cupelled.

The weighed ore and fluxes are mixed in the capsule with the handle of the little spoon, and by means of a hair brush or spatula introduced into a soda-paper tube, closed at one end. The open end is folded and the tube placed in a cylindrical hole in a piece of charcoal, made with a coal borer.

b. The Fusion.—The paper is first burnt off with a feeble flame, then the whole test covered with a strong reduction flame, which must continue till slag and lead appear separated. The slag must be free of lead globules. Under these circumstances all the silver is taken up by the lead, and the earthy matters by the borax. The fusion is now finished. But it is generally the case that different volatile substances are combined with the lead; for instance, sulphur, antimony, etc., which must be driven out by longer blowing with the oxydation flame, directed on the lead button. This operation must continue till the lead appears white when cold. It is then taken out with steel forceps and freed from adhering slag on the steel anvil, giving it a cubic shape.

c. First Cupelling.—The cupel is prepared from sifted boneash, which is beaten in a little iron mould by gently striking the pestle with the hammer. This iron mould is placed on a small wooden stand and glowed with the flame. When this is done; the lead button is laid on the cupel and melted with the blue flame, which must touch the lead till it begins to operate. The blue, sharp pointed flame is then kept a little further off, being always directed on the button, which must be kept at a moderate red heat, so that the rising litharge may congeal on the sides. If the heat is too strong, the lead evaporates, and the litharge draws into the cupel-mass, or fuses over the test. If, on the other hand, the temperature is too low, the litharge accumulates too much on the lead, covering the latter, and stopping the operation. When the lead button is reduced to the size of a hemp seed, the blowing must be stopped, the button seized with a pair of pincers, freed from boneash and flattened somewhat.

d. Fine Cupelling.—The used boneash is removed from

the iron mould, and another test beaten in, with a smooth surface, which is best obtained, if levigated boneash is mixed with the sifted. After the cupel has been dried by the oxydation flame, the flattened lead is placed on the side of the test and exposed to the oxydation flame. In melting, it assumes the globular form. The cupel must be held inclined, and slowly turned, so that the button can slide on the surface of the test. The flame is directed always around the lead, and so much heat applied, that the button appears red-hot and the litharge disappears in the cupel-mass. When the button assumes the greenish color of fused silver, the operation is finished.

e. The Weighing.—The weight of the silver globule is determined either by weighing on the balance with gramme or grain weights, or, if too small, by measuring it on a scale. The scale is engraved on polished ivory. There are two convergent lines forming an acute angle. These lines are divided into equal sections by parallel lines, crossing the convergent lines. On the right side, each section shows the per centage of silver. The round or spherical silver globule is placed between the convergent lines and moved down, till it becomes tangented by the convergent lines. The numbers on the right side show then the amount of silver. This measuring must be done with the aid of the magnifying glass.

2. Argentiferous Alloys.

Alloys rich in silver, for instance, native silver, amalgam silver, silver coin, etc., are dressed with one grain of test lead, and twenty-five per cent. of borax glass to one grain of alloy. Alloys containing more copper must be fused with from two to five grains of lead. This is done in a groove on the charcoal with the reduction flame, till lead and alloy are melted into one mass and no globules are perceived in the slag. It is then cupelled and weighed.

Poor substances can be cupelled directly, if suitable for cupellation, for instance, pig lead, or argentiferous bismuth. Of these substances, two to five grains may be weighed out for cupellation. Copper and nickel containing alloys (block copper, German silver, brass, copper coin) must be weighed out in portions of one-half grains, and fused with ten grains of lead and one-half grain of borax glass, in a soda paper tube, with the reduction flame. The obtained button is cupelled. In the second, or fine cupellation, some lead may be added, if a great deal of copper is in the alloy. If there is also tin besides the copper in the composition, one grain of it must be fused with five to fifteen grains of lead, one-half grain of borax, and one-half grain of soda, in the reduction flame, finishing the operation with the oxydation flame. If the lead does not exhibit a whitish color, it may be melted for a while with the oxydation flame on another spot of the charcoal with borax, and then cupelled. Argentiferous quicksilver combinations are treated first in a closed glass tube with the alcohol flame, in order to drive off the quicksilver, then melting the residue of the one grain test, according to the amount of copper, with one to three grains of lead and some borax. If the residue adheres closely to the glass, it may be fused with the cut part of the tube, lead, and soda. Argentiferous iron and steel are fused, one grain of either, with one grain sulphur, eight grains lead, and one-half grain borax, in a paper tube. They must be treated with a good oxydation flame until all sulphur is separated, and the button shows a lead color. A combination of silver with tellurium, antimony, and zinc, is melted with five grains of lead and some borax, with the reduction, and finally with the oxydation flame.

CHAPTER II.

METHODS OF EXTRACTING SILVER.

Sec. 65. The extraction of silver requires generally complicated processes, as it occurs in most cases combined with other minerals which must be removed before the silver can be obtained. Generally there are three methods in use by which the silver is extracted, the choice of which depends on the nature of the ore: 1. Melting of silver ores with lead; 2. Amalgamation, and 3. Dissolving and precipitating the silver.

1. *In Melting Silver* **Ores** *with* **Lead,** *Oxyd of* **Lead,** *or* **Lead** *Ores*—silver containing lead (pig lead) is produced, which is separated directly by cupellation, or if the lead is poor in silver, the latter is concentrated in the lead by Pattinson's method. The advantage of this procedure depends principally on the price of fuel and lead, if the latter is not present in the ore. The greatest part of the produced silver is extracted from argentiferous lead and copper ores, the smaller portion of real silver ores. The extraction of silver from argentiferous galena is one of the oldest processes which, by its simplicity, is now

conducted almost in the same way as it may have been thousands of years ago. The extraction of silver from argentiferous copper ores, requiring more advanced metallurgical knowledge, is a later process. The extraction of silver may be executed on copper ores or on its melting products (copper matt or black copper), but this process must always be considered very imperfect, on account of the considerable loss of silver, lead, and copper. The affinity of these metals for each other makes the process complicated and expensive. Copper ores not rich in silver are subjected to melting by which the silver is concentrated in the obtained black copper, this being separated from silver by liquation. Copper ores rich in silver are melted with lead or lead ores, by which a suitable copper is obtained for the liquation. To avoid this liquation process, repeated roasting and melting of copper matt with lead containing substances is adopted, but the desilverization of the matt is imperfect.

Although the behavior of silver, copper, and lead towards the sulphur, on which this process is principally based, is well understood, the imperfections are not yet removed. For this reason more attention is now paid to the working of argentiferous copper ores by amalgamation and precipitation processes.

It is preferable to add rich silver ores to the lead in the cupelling furnace, poor silver ores bare of such substances as collect the silver from the slag, are fused with iron pyrites, if no lead ore can be obtained. The earthy substances scorify and the silver is taken up by the sulphide of iron (iron matt) which must be treated afterwards with lead. Most of the silver ores are copperiferous to some degree, yielding, also, copperiferous matt, of which the silver, by means of lead, cannot be extracted perfectly. The copper makes the process complicated. In melting, the scorification of the silver is prevented by the great affinity of sulphur to silver.

According to Karsten, poor copperiferous silver ores can be worked advantageously for silver, under favorable circumstances, when the ore contains only ten ounces of silver to the ton. But also six and even three ounces per ton will pay the extraction, if copper or lead is obtained from the same ore. The value of silver is eighty times higher than that of copper, and four hundred times higher than lead.

The silver containing lead, obtained in either way, is parted by an oxydation process more than a thousand years old, called cupellation. If the lead is poor in silver, it is first subjected to the Pattinson's crystallization process, by which the silver is concentrated in the lead. Parke's method of extracting the silver from the lead by zinc, promising great advantages, is not yet extensively practiced, on account of some of the zinc remaining in the lead.

2. *The Amalgamation or Extraction of Silver by Quicksilver in the Wet Way*—is based on the facility with which mercury unites with the silver, forming amalgam. It is easily separated by glowing. This process is said to

have been discovered about the middle of the sixteenth century in Mexico, and became known in Europe in the latter part of the seventeenth century.

This process is preferable to the preceding, when fuel is expensive, or in treating poor silver ores, argentiferous copper ores, or copperous products, free of lead. Of poor, earthy silver ores, the silver can be concentrated in sulphide of iron (iron matt) by melting with iron pyrites. Only the real silver ores are subjected directly to amalgamation. From argentiferous copper ores the silver is concentrated in the copper matt, or in black copper; these products are then amalgamated, allowing thus a cheaper manipulation, a more perfect silver extraction, less loss in copper, and the production of purer copper, than the melting process. The European amalgamation has the advantage over the American (patio) in yielding more silver, there being considerably smaller loss in quicksilver, while the patio amalgamation, on the other hand, can be carried on without buildings and fuel.

3. *Solution and* **Precipitation** *of Silver.*—There are principally two processes, lately invented by Augustin and Ziervogel, which are superior to amalgamation, being simple, cheap, and extracting the silver in a short time. Augustin transforms the silver into a chloride, by a chloridizing roasting, dissolves the chloride in concentrated solution of common salt, and precipitates the silver by copper. Ziervogel roasts the ore, under conditions

that the silver is converted into a sulphate, which is **soluble** in hot water and can be precipitated by **copper,** after lixiviation. This last method is more simple, **but** leaves richer tailings than the preceding. Both processes give, then, only a good result, if the roasting is conducted with the utmost care, and if the ore or matt **is free** from **lead,** zinc, antimony, and arsenic, **or,** at **least, if** only **small** quantities of these injurious metals are present. **The use** of **these two processes, therefore, is** limited to places where **pure ore is** found, or pure matt obtained.

Other extracting methods of this class, for instance, by **common salt and** ammoniac, salt and chloride of copper, **etc.,** have **to be** considered as experiments rather than established processes.

According to the preceding, **the most** suitable way of treating the following different **silver ores will** be:

1. *Real Silver Ores*—With earthy and sulphuretted substances, with pyrites, blend, **and** copperous ores, must be treated according to their richness:

a. Rich or middling rich ores containing **fifty** to ninety per cent. of silver, are best subjected directly to cupellation with lead. Forty per cent. or less silver containing ores can be melted with **lead ores and** fluxes in a vertical blast furnace. **Such** ore cannot **be** worked to advantage **by** amalgamation (except in pans, as already described). **Very** rich ores, especially if they contain native **silver,** are **melted also in crucibles** with proper **fluxes.**

b. Poor ores of this class, containing from two and a half to forty ounces per ton, are subjected to melting with iron pyrites (melting for matt). The silver of the obtained matt can be extracted either by lead or by quicksilver. The extraction with lead is proportionately more perfect the less quantity of copper there is in the matt. Ore with from forty to sixty ounces of silver per ton can be amalgamated with a proper addition of iron pyrites in roasting.

2. **Pyritous ores,** containing silver ores disseminated in iron pyrites, yielding sufficient matt in melting, are subjected to such melting in blast furnaces, and the obtained matt treated with quicksilver or lead. Richer in silver, this class of ore is best suited for the amalgamation, but the presence of copper pyrites, arsenical pyrites, zinc blend, or galena, is not favorable. Richer ore may be also melted with lead containing ores after roasting.

3. **Argentiferous lead ores are entirely** unsuitable for amalgamation. Their beneficiation is best effected by melting, and it depends, further, on the choice of the most suitable desilverization of the different products. The less copper in the ore, the easier and more perfect the extraction of silver and lead.

4. Argentiferous copper ores can be treated in different ways:

a. The poorer ore is subjected to melting for **matt, or black copper**. From these products, the silver is then extracted. The copper matt is desilvered either by lead or quicksilver, or by Augustin's or Ziervogel's methods. Argentiferous **black** copper is oftener **treated with** lead than **with** quicksilver. The direct amalgamation of argentiferous copper ores is often disadvantageous.

b. Richer copper ores are **melted with proper dressing** in roasted or **unroasted** condition, with roasted or unroasted lead **ores.**

c. **Very poor** copper ores are added **to the melting** for iron **matt (melting with iron** pyrites).

5. Argentiferous **blend and arsenical pyrites must be** roasted, then subjected **to melting for iron matt.** In roasting such **ores, a** considerable **loss of** silver may **occur.** If melted without roasting, **the** blend causes a matt fusible **with** difficulty, by which the silver is not perfectly collected. Argentiferous arsenic can be **melted with** lead after careful roasting.

6. Argentiferous cobalt **ores are beneficiated, first for** smalt, the **silver, then** extracted from **the** residue by lead or amalgamation.

Sec. 66. The principal known methods **of extracting silver can be brought into the** following **division:**

EXTRACTION OF SILVER IN THE DRY WAY.

A. LEAD MANIPULATION WITH ARGENTIFEROUS ORES.

a. Melting of rich silver ores in crucibles with lead or litharge.

b. Fusion of rich silver ores with lead. (Cupellation.)

c. Melting of rich unroasted silver ores, with roasted or unroasted lead ores, or plumbiferous products, cast iron, lime, or iron slag (virtuous slag) in blast furnaces.

d. Melting of argentiferous lead glance in unroasted condition, with desulphurating substances (metallic iron, iron ore, ferriferous matts, etc.).

e. Melting of roasted argentiferous lead glance with desulphurating substances (cast iron, roasted iron matt, and iron slag).

f. Melting of pure unroasted lead ores without desulphurating agents. (Melting in North-American lead hearths).

g. Melting of roasted argentiferous lead ores without desulphurating agents in blast furnaces, in reverberatory furnaces, or in hearth furnaces.

h. Melting of roasted argentiferous copper ores with roasted lead ores, or with products of cupellation.

B. LEAD MANIPULATION WITH ARGENTIFEROUS MATTS

Which are obtained, either in melting of argentiferous ores with iron pyrites, or in melting of argentiferous lead and copper ores:

a. Treatment of the unroasted iron, or copper matt with metallic lead:

1. *Fusion.* The melted matt is mixed with melted lead.

2. *Hydrostatic Melting.* The melted matt is forced through a body of fused lead.

b. Melting of unroasted copper matt with plumbiferous products.

c. Melting of roasted iron matt with roasted or unroasted lead ores or with products of cupellation.

d. Melting of unroasted lead matt with unroasted galena and iron.

e. Melting of roasted lead matt with iron.

f. Melting of roasted lead matt without desulphurating agents.

C. MELTING LEAD WITH ARGENTIFEROUS BLACK COPPER (LIQUATION.)

D. EXTRACTION OF SILVER WITH COPPER AND LEAD.

Argentiferous copper matt is melted with black copper, spongy copper from liquation, and plumbiferous products, whereby a part of the silver is disengaged by the copper and taken up by the lead.

E. SEPARATION OF SILVER FROM ARGENTIFEROUS LEAD.

a. Cupelling in furnaces with a fixed hearth.

b. Cupelling in furnaces with a movable hearth.

F. CONCENTRATION OF SILVER IN PIG LEAD BY PATTINSON'S CRYSTALLIZATION PROCESS.

G. REFINING OF SILVER OBTAINED FROM CUPELLATION.

a. Refining in ovens with fixed or movable tests:

1. Under the muffle.
2. Before the bellows.
3. In reverberatory furnaces with movable or immovable tests.

b. Refining in crucibles with or without agents.

H. DESILVERIZATION OF LEAD BY ZINC.

EXTRACTION OF SILVER IN THE WET WAY.

A. EXTRACTION BY QUICKSILVER. (AMALGAMATION).

a. European barrel amalgamation.

b. Amalgamation of ores.

c. Amalgamation of copper matt.

PROCESSES OF SILVER AND GOLD EXTRACTION.   215

    d. Amalgamation of speiss (arsenides, ger. speise).

    e. Amalgamation of black copper.

    f. American heap amalgamation. (Patio.)

    g. Combined American and European amalgamation.

B. EXTRACTION BY SOLUTION AND PRECIPITATION. (AUGUSTIN'S METHOD).

    a. Extraction of silver from matts.

    b. Extraction of silver from ores.

    c. Extraction of silver from speiss (ger. speise).

    d. Extraction of silver from black copper.

C. ZIERVOGEL'S WATER LIXIVIATION ON ARGENTIFEROUS COPPER MATT.

D. PATERA'S METHOD OF EXTRACTION OF SILVER FROM ORES.

CHAPTER III.

EXTRACTION OF SILVER IN THE DRY WAY.

Extraction of Silver with Lead.

SEC. 67. The use of lead for the extraction of silver is based:

1. On the property of the lead to decompose the sulphide of silver under formation of sulphide of lead. If there are other sulphuretted metals in the silver ore, especially iron and copper sulphurets, they will be less decomposed by the lead.

The products of melting are argentiferous lead and matt of iron, copper, and lead. The less copper there is in the ore, the easier and more perfect the extraction of silver, because the sulphide of copper retains a part of the sulphide of silver in the matt with great obstinacy, and causes a repetition of the extraction. The silver ores are very seldom free of argentiferous copper ores; a copperous matt is therefore always obtained, the silver of which never can be extracted perfectly by lead or lead ores. This portion of the silver is therefore given up entirely, or extracted by liquation, when black copper is obtained from the matt. The sulphides of iron and lead also retain a small amount of silver.

This chemical process takes place in melting rich silver ore with lead crucibles, with lead in a cupelling furnace or in treating argentiferous matt with metallic lead, also partly in melting argentiferous lead glance with metallic iron.

2. On the decomposing action of oxyd of lead and of sulphate of lead on sulphuret of silver in such a way that argentiferous lead and sulphurous acid are formed. This process occurs if unroasted sulphuretted silver ores or argentiferous matts are melted with roasted lead ores or oxydized lead combinations. The reduced lead acts also on the sulphide of silver in the manner mentioned under 1. In this mode of melting also the appearance of matt with some silver cannot be avoided, inasmuch as some of the undecomposed roasted lead ore and the sulphides, contained in the silver ore, also the sulphides of reduced sulphates form argentiferous matt.

3. On the reductive action of lead on oxyd of silver or on sulphate of silver. This reaction takes place in cases mentioned under 1 and 2, if roasted silver ores or matts come in contact with lead. Sulphuret of silver, roasted by itself can be transformed perfectly into metallic silver, emitting sulphurous acid, but combined with other sulphurets, for instance with sulphurets of iron or copper, a part of sulphate of silver is always formed,

which decomposes at a higher temperature than the sulphates of iron and copper.

4. On the greater affinity of silver for lead, than copper. A fusion of argentiferous copper and lead produces a compound of argentiferous lead and a more difficultly fusible alloy of about three parts copper and one part of lead with a small amount of silver. The argentiferous lead can be separated from the copper by liquation.

The extraction of silver with lead is an imperfect process, it suffers a great loss in silver, copper, and lead, is complicated and expensive and furnishes a copper of low quality. In place of this mode the more perfect amalgamation process is adopted, which again commences to give way to the still simpler and cheaper silver extracting processes of Augustin and Ziervogel. The choice of beneficiating methods does not depend on the nature of the ore alone, but also on the local circumstances, for instance, cheap lead ores, etc. Whether the desilverization must be adopted directly with the ore, or with the matt or black copper, resulting from the ore, depends chiefly on the amount of silver and local circumstances, also from the quantity and price of the lead ores.

The loss of lead in melting plumbiferous ores arises principally by scorification and volatilization. The scorification is produced either by a chemical combination of oxyd of lead with the compounds of the slag, or in a mechanical way by the adhering of lead or matt to the slag. The remedy for the first scorification can be obtained only by a proper dressing of the ore, for the latter principally, by giving a greater depth to the furnace from the breast wall to the back wall. In this case the slag is bound to take a longer time in flowing out, thus having more chance to deposit the lead globules.

The volatilization of lead depends on the speed of the blast, which rises in the melting room of the furnace, and on the pressure of the heated gases. The more violent the wind presses upwards the more the lead volatilizes. This occurs mostly near the nose (slag channel) at the back wall, where the ore is impregnated with finely-parted metallic lead. In making the furnace deeper from the breast wall towards the back wall in the direction of the wind, contracting the space above the melting region and giving the furnace a sufficient height, the loss of lead by volatilization will be diminished. The height and depth of the furnace depends on the quantity and pressure of the wind required for the melting. The furnace must be deeper and higher, if more wind and pressure are necessary. The best remedies to diminish the loss of lead therefore, are: the height of the furnace, the enlargement of the melting room in the direction of the wind, and the contraction above the tuyere.

PRODUCTION OF ARGENTIFEROUS LEAD.

Lead Manipulation with Argentiferous Silver Ores.

Sec. 68. To this treatment are subjected rich real silver ores, rich roasted **pyritous ores,** argentiferous-lead and copper ores in **roasted** and unroasted condition, etc. As plumbiferous substances are used: metallic lead, roasted or unroasted lead **ores and the products** of cupellation (litharge, heath).

Melting of Rich Silver Ores in Crucibles.

Rich silver ores, are worked in the cheapest and most **perfect manner** when added to cupellation. Where the cupelling process is not in use, the ore is melted **in** crucibles (either black lead or clay), generally with an addition of iron to decompose the sulphide of silver, and **with some lead if there is none in the ore.**

If the ore is free from **earthy** substances, some potash and powdered glass are **added,** but if earths are present it is necessary **to use** some litharge, which forms a better slag. The plumbiferous-obtained silver is then refined. If there be some lead and silver in the matt **it** is **melted** again with an addition of iron.

The silver ore from Kongsberg, Norway, consists chiefly of native silver, silver glance, brittle silver ore, ruby silver, **antimonial silver** and lead glance accompanied by iron **pyrites, zinc** blend, copper pyrites and **magnetic** pyrites. The **richest ore, from** ten to thirty

PROCESSES OF SILVER AND GOLD EXTRACTION.   221

per cent. in silver, **is** treated in crucibles. There is obtained by this process the largest part of the Kongsberg silver production. The melting is performed in **clay** crucibles in a draft furnace, in quantities of three **hundred** pounds. For the scorification **of** iron some **quartz and one** per cent. of borax are added. The same **crucible** is charged from four to six times daily **and** the following **products obtained**:

a. Silver which is refined on movable tests **of wood** ashes and lime, **with an** addition of half **an ounce lead to** twenty-four ounces **of** silver. The refining **of eight hundred ounces** of silver consumes one hundred **and** ten cubic feet of pine charcoal. The average loss of silver in **refining** amounts **to about** seventeen **per** cent.

b. Slag which is pulverized, the coarser silver buttons extracted and **the** balance turned over to the smelting furnace.

Fusion of rich Silver Ores with Lead in Cupelling Pig Lead.

When the lead is brought so far in operation that litharge commences to separate, the wind is stopped and the rich ore introduced **on the** surface of the lead bath by means of a scoop, in quantities of from one hundred to two hundred pounds, according to the size of the hesth. Then the heat is raised for about **one hour.** During the time the ore is roasting the **silver draws** into the lead, the mass **comes** into a very fluid condition and the blast is let on again. The slag which contains all the earthy substances of the ore is drawn off with a wooden pole, and on account of some retained silver is given over to the smelting furnace. To this process the ore is subjected which contains 3,000 or more ounces of silver per ton. In China, roasted ores are used for this method.

Melting of Unroasted Rich Silver Ores with Unroasted Lead Ores or Plumbiferous Products.

This method, in regard to silver extraction, is advantageous under circumstances. Lead ore or lead-containing products must be added in sufficient quantity to fully cover the silver, otherwise a considerable loss of silver may occur by scorification. This loss increases if low furnaces are in use, and if the dressed ore is hard to melt. If in want of lead ores or lead-containing products, it is then not advisable to subject too rich silver ores to melting with iron pyrites, concentrating thus the silver in the iron matt.

At Allemont (Isère department) the silver ores (silver glance and rubysilver) are dressed in the following proportions: one hundred tons of ore with about two hundred and forty ounces of silver to the ton, are dressed with one hundred and fifty tons of slag, sixteen tons of quicklime, sixteen tons of iron slag and as much galena, litharge, and hearth as is necessary to obtain pig lead, with about six hundred and forty ounces of silver per ton. The galena is decomposed by the iron of the ore and by the iron slag. The melting is performed with charcoal in furnaces three and a half feet high. The result is not perfect on account of the low furnaces and defective dressing of the ore.

At St. Andreasberg, the furnaces are twenty-two feet high, and so much lead ore is taken that half an ounce of silver is protected by seven pounds of lead. The ore is dressed: thirty-eight tons of ore, fifty tons slag, four to six tons metallic iron, and forty-eight tons lead-containing substances. The obtained pig-lead contains from one hundred and sixty to two hundred and forty ounces of silver. The total loss of silver in products and by volatilization is from 4·43 to 4·5 per cent.

MELTING OF ARGENTIFEROUS LEAD ORES IN ROASTED AND UNROASTED CONDITION, GENERALLY WITH SOME ARGENTIFEROUS COPPER ORES.

To this heading belongs several lead-melting processes. Almost all galena contains some silver which after melting is found chiefly in the lead and partly in the matt. The copper of the ore concentrates in the matt. The silver in the matt is in small quantity and either not regarded at all, or the roasted matt is melted either by itself or with plumbiferous substances, concentrating thus the silver in the lead, or the silver is extracted from the black copper obtained from this matt.

Melting of Roasted Argentiferous Copper Ores with Roasted Lead Ores.

This method is advantageous, if the copper ores are not rich in silver. If the amount of silver is too large, it is exposed to scorification. In this case, the ore must be melted without roasting and the silver collected in the matt, which for the purpose of desilverization can be treated in different ways, as mentioned further on. Melting processes, not melting for matts, cause a greater loss in metals, because the basis for scorification in melting is laid in the preceding roasting, which is not the case with unroasted ore, where only a separation of earths and sulphurets takes place.

The argentiferous copper ores are slightly roasted and melted with perfect roasted lead ores or products from cupellation in low furnaces. The obtained pig lead must be cupelled. The matt is roasted over several times, and melted with the products of cupellation. After the third melting with lead the matt is generally free from silver, but the matt from rich ore usually retains silver to considerable extent. After roasting, it is melted for black copper and the silver extracted by liquation. This process is expensive and imperfect, on account of losses in silver, copper and lead.

LEAD MANIPULATION WITH ARGENTIFEROUS MATTS.

SEC. 69. Poor silver ores do not suit for direct melting with lead ores. They produce a tough melting, increase the scorification of lead, etc. Such ores are first subjected to melting for matt, that is, they are melted with iron pyrites. The earths and useless oxyds form the slag, while the silver is concentrated in the matt. This iron matt, unlike ore, assists the subsequent melting with lead ores, and protects a great part of the lead against scorification; it also acts as a precipitation agent for lead by which the expense of the melting for matt is lessened. Another advantage of this method is the beneficiation by the way of all rubbish and slags of rich meltings of which the silver, lead, and copper are taken up.

I. CONCENTRATION OF SILVER IN MATT.

The melting for matt is called in German raw work (Roharbeit), because the ores are subjected to this process raw, without preparation. The object is, as before mentioned, to concentrate the silver in a small quantity of matt or sulphides (Germ. Rohstein), by means of iron pyrites or heavy spar (if the ore does not contain already a sufficient quantity of sulphur), while the earthy substances and foreign oxyds of metals are scorified. In melting for matt the following rules must be observed:

1. The mixture of ore and fluxes must be prepared so that:

a. A certain quantity of matt is obtained in proportion to the amount of silver in it.

b. A proper fluid slag is formed.

In regard to the proportion of matt to the amount of silver, the concentration must be limited from forty to sixty ounces per ton, according to experience, otherwise a considerable loss of silver might occur. The loss is a mechanical one which cannot be avoided, caused by small particles of matt becoming involved in the slag. To produce a rich matt, it requires a smaller quantity of iron pyrites in consequence of which the silver cannot be collected perfectly remaining exposed to scorification.

The most suitable amount of matt in the ore is between thirty and fifty per cent. of the prepared ore. This amount must be obtained by a proper proportion of pyritous ore or by addition of iron pyrites. If the ore is found to be too rich in pyrites, some sulphur must be driven out by roasting. Some copper pyrites is advantageous, it imparts to the iron pyrites a greater collective ability and separates better from the slag; but a copperous matt is more difficult to treat with lead for the purpose of silver extraction. Arsenical pyrites and zinc blend, if in considerable quantity, are injurious, weakening the collective power of the matt. The first causes a loss of silver by its volatility, the second produces a tough melting.

The quantity of matt in the prepared ore is ascertained by the assay for matt. For this purpose, one-half ounce of borax glass, one-half ounce powdered glass, twenty-four grains of resin, well mixed with one-half ounce of ore are introduced into a crucible and covered with one and one-half ounces of common salt. This mixture is melted for about half an hour, applying a strong heat. The matt button obtained must be weighed directly, as it falls to powder after some time.

If substances with a small amount of sulphur are assayed for matt, some metallic copper must be added to the above mixture. The regulus separates easily from the slag. The added copper must be subtracted from the weight of the regulus.

As mentioned under *b*, care must be taken to obtain a fluid slag in melting the ore, in order to facilitate the separation of matt. This depends on the right proportion of the substances producing the slag. Ores abounding in earthy substances, are prepared with ferriferous ores or roasted iron pyrites; some lime is also added, especially if a great deal of quartz is present. Iron slag gives also a good flux. A favorable scorifying substance is found in the single silicate slag, resulting from melting lead ores. The melting for matt is in the best condition, if a slag is obtained between bi- and singulo silicate, which permits an easy separation of the matt, without stiffening too quick. This mode of melting is more difficult if, in absence of iron pyrites, ponderous spar is used.

The melting is performed with charcoal or coke, in either blast or reverberating furnaces. The blast furnaces are provided with one, two, or more tuyeres, and with dust chambers. A furnace with two tuyeres melts more ore than with one, and consumes also less fuel. Reverberating furnaces have several advantages over the blast furnaces, but yield a richer matt in zinc, when blend containing ore is worked. This, however, can be remedied with an addition of charcoal and roasted ore containing oxyd of iron. Heated compressed air favors the melting and effects saving of fuel.

II. EXTRACTION OF SILVER FROM IRON, COPPER, AND LEAD MATTS BY MEANS OF LEAD.

The desilverization of the matts can be performed:

1. Treating the matt in unroasted condition with metallic lead, or substances containing oxyd of lead, or with galena and metallic iron.

2. By melting the roasted matt with roasted or unroasted lead ores, litharge, and hearth.

3. By melting of roasted lead matt either by itself or with metallic iron.

Treatment of unroasted matt with metallic lead, is

based on the nature of metallic lead to decompose the sulphides of silver, forming sulphide of lead. The separated silver is taken up by the surplus lead. The desilverization is obtained more perfectly if a better contact between lead and matt is effected, but a perfect extraction cannot be accomplished, unless by repeated melting with fresh lead. In this case, the black copper, obtained from the matt is not subjected to the liquation. This procedure is generally adopted where no lead ore can be got. In melting the matt with lead ores in blast furnaces, the desilverization of the matt is more perfect, but a greater loss in lead and copper will result, than by treating the matt with metallic lead at a lower temperature.

There are only two ways for the treatment of matt with lead—The hydrostatic melting, and melting of lead with matt:

A. *The Hydrostatic Melting*—or melting through a column of lead. The operation will succed, if the matt is conducted through the lead by drops. The matt must press on the lead with greater weight than the weight of the lead itself. It is therefore necessary to have the tuyere at the proper height above the front hearth. To find the pressing height of the melted masses (matt and lead) we must observe, for example, the melted matt in the furnace $= a$, the specific gravity of the matt $= b$, and the gravity of the lead $= c$. If, for instance, the height of the copper matt column $a = 30$ inches, $b = 5\cdot 0$, and $c = 11\cdot 33$, and x, the desired height of the front hearth, which is to be filled with lead, the answer would be:

$$x : a = b : c, \text{ or,}$$
$$x = \frac{a \times b}{c} = \frac{30 \times 5\cdot 0}{11\cdot 33} = 13\cdot 2 \text{ inches.}$$

Plumbiferous matt is less suitable for this operation than matt containing arsenic and antimony. Generally, this hydrostatic melting with suitable copper matt, compared with the liquation, suffers a smaller loss in copper and lead, but the desilverization is not so perfect. In deciding for the preference of these two modes, the question must be considered whether the value of the greater quantity of obtained lead and copper covers the value of the less quantity of extracted silver.

B. *The Fusion of Lead and Matt.*—The fused matt is stirred and mixed with the melted lead in a hearth prepared for tapping. The hearths are of different sizes, about four and one-half feet in diameter, and three feet deep. The hearth is provided with a pair of bellows. The matt is first melted down on charcoal, then the lead, which sinks through the matt to the bottom, taking up silver from the matt. Four hundred pounds of matt are treated with about seventy-five pounds of lead. When all the lead is melted, it must be mixed for a while with the matt by means of an unseasoned wooden rod, and the lead tapped. The matt must be treated with lead repeatedly, three or four times over. The lead of the first tapping is ready for cupellation, but that from the following is used over again for the extraction, till the silver concentrates to about three hundred ounces per ton, when it is considered rich enough to be cupelled.

LEAD MANIPULATION WITH ARGENTIFEROUS BLACK COPPER (LIQUATION).

SEC. 70. *Argentiferous Black Copper*—Is produced, if the copper ores, or the matts are so poor in silver that the treatment with lead is not advantageous. In this case the silver is purposely concentrated in black copper. Argentiferous black copper is also obtained from melting of argentiferous galena containing some copper ore, or from working argentiferous copper ores, or matts with lead. Although the greatest part of the silver is extracted from these substances by the lead, the copper always retains some silver as long as sulphur is present.

The separation of silver from copper can be executed in different ways. One of the oldest methods is the liquation. Argentiferous copper is melted with a certain proportion of lead. In cooling, a great part of the silver containing lead separates, forming then a mixture of argentiferous lead and an alloy of about one part lead with three parts copper, and considerably less silver. Heated to the melting point of lead, the latter, with the silver, flow out, leaving the alloy of copper and one part of lead in unmelted spongy condition, retaining its original shape.

This method is defective, not permitting a perfect separation of silver, and suffers a great loss in silver, copper, and lead. A great many products result from this operation, so that the liquation process appears endless. For this reason, in most places, other processes are now adopted, by which the silver from the matts is extracted in a more perfect way. In Freiberg (Saxony) Augustin's method is adopted. In Mansfeld, the liquation was replaced, first, by amalgamation of the copper matt, then by Augustin's, and finally by Ziervogel's extraction. According to Karsten, the cost for liquation of one hundred pounds of black copper, is equal to the value of four or four and a half ounces of silver, so that black copper with eighty or ninety ounces per ton, would not pay the liquation. This is the reason that some copper of commerce contains a great deal of silver. In some places, however, copper with forty-five to fifty ounces of silver per ton, was subjected to liquation with profit.

But the expense of liquation increases, not only at a very low amount of silver, but also when the copper is rich. The reason is that in this case, if the copper contains, for instance, twenty per cent. of silver, the liquation must be often repeated, in order to extract as much silver as possible. The operation becomes extended, so that the loss of silver must increase. The cost of liquation of one hundred pounds of copper will be more than the value of the four and a half ounces of silver.

The black copper is first broken in small pieces cold, or it is made red-hot, which facilitates the breaking, or it is melted and poured into cold water (granulated), and then melted with lead. Each ounce of silver requires thirty to thirty-two pounds of lead, but the proportion of lead to copper **should not be** over 11 : 3. Lead and copper **in the proportion** of eleven **to three,** in regard to the above mentioned proportion of lead and silver, are melted in a cupola furnace with aid of **the bellows.** The alloy is poured into moulds, assuming the shape of a disk eighteen inches in diameter, and from three **to three and** a half inches thick.

These copper disks are **laid on two** inclined iron plates in an oven, **in such a way that a space of one or two** inches is left **between each disk.** They are covered with charcoal, **and it ignited.** The lead soon commences to trickle **from the disks, and runs on** the inclined iron plates, through the split **which is left between** them into a basin underneath. When **the lead** ceases to flow, the operation is finished. **The lead is given** over to the cupellation. **The** spongy copper contains from ten **to** twenty per cent. of lead. If this copper is found to be rich in silver, it is fused again with lead and treated as before. If otherwise, it is subjected to the sweating process.

The sweating is performed **in another** furnace at a stronger heat, **to get rid of** another portion of lead, which however **is obtained in an** oxydized condition. **The** unfused copper is purified on a refining hearth.

Extraction of Silver by Copper and Lead.

The object of this process is the decomposition of the

sulphides of silver in the ores and matts, combined with an extraction of silver by lead. The decomposition of silver sulphurets by copper is not perfect for the reason that when copper prevails, an alloy of silver, copper, and matt is formed which retains a great deal of sulphide of silver. If there is also lead in the ore or in the melting product, this will be eliminated with the silver while an equivalent quantity of copper enters the matt.

Copper-Dissolving Process.

The principle of this process is as follows: If argentiferous matt is melted together with copper and plumbiferous products, the copper enters the matt, while the eliminated silver is taken up by the lead. At the same time the lead acts, desilvering the matt. If too much copper is added it does not remain dissolved in the matt, but separates partly again, and enters the lead. If the copper contains silver, this will be also extracted, so that the copper need not be subjected to another extraction, for instance, by liquation or amalgamation. The usefulness of this process depends generally on local circumstances, especially on the proportion of matt and black copper, which is obtained in melting. If too much matt, then the extraction of silver is imperfect; on the contrary, if too little, the black copper cannot all be dissolved by the matt, but alloys with the lead.

Under suitable circumstances, this process is advantageous. It is preferable to the liquation process, because it allows a more perfect desilverization of the matts.

SEPARATION OF SILVER FROM ARGENTIFEROUS LEAD.

Cupellation of Pig Lead.

Sec. 71. Cupellation, the object of separating silver from argentiferous lead, is executed in such a way that the lead on the hearth of an oven, under influence of heat and compressed air, is converted by degrees into an oxyd of lead (litharge) which draws to the periphery from the convex surface of the fused lead, exposing thus the metal constantly to the oxygen of the air. Drawing off this litharge as it is formed, the silver finally remains in the hearth.

The old cupelling furnace has no separate fire-place, but the lead was melted in the hearth between firewood; the blast was directed on the lead, over which burning wood was maintained till the process was finished. This mode requires much more fuel and the oxydation of lead is imperfect.

There are two modes of cupelling in use: The German on fixed hearths, and the English on movable tests.

Cupellation in unmovable Hearths.—The German cupelling-furnaces are constructed of two principal parts; of the fire-room and of the cupelling-room, connected by a fire-bridge. The cupelling room is formed by a circular block of mason-work about fifteen inches above the ground, provided with channels for the escape of moisture. On the periphery of the block, the wall from twelve to fifteen inches thick is carried up twenty inches, forming a wall-ring, of which the inner space receives the material for the hearth. The coppel or hood is formed of many iron bars well riveted, and with many little hooks for the purpose of holding the clay lining. There have been also clay coppels in use, weighing over a ton; the iron ones, from 1,100 to 1,300 pounds, are generally preferable for many reasons. By means of a crane, three coppels can be lifted and turned aside. Arches of brick, used in some places, are less convenient. They must form high coppels (five and a half feet) to resist the effect of heat and lead fumes longer; they are a great deal cheaper, but the iron ones stand ten times as long. The iron skeleton of the coppel is lined first with loam, then with a mixture two inches thick of three parts of clay and one part of quartz.

In the wall-ring, coarse slag is introduced or rock of the size of one's fist, about twelve inches deep in the centre, and sixteen to eighteen inches deep on the sides, forming a concave surface. Over the slag comes a row of brick and then the hearth-mass. Lixiviated wood-ashes are not much in use, but replaced by the marl which is a much better hearth material. The marl is cheaper, it does not absorb so much litharge, and the result in regard to the quantity of silver is always better than on a test of ashes. It is pulverized and sifted through a sieve of about sixty-four holes to the square inch, then very evenly moistened, introduced in the hearth and stamped hard from four to six inches thick. The whole mass may be introduced at once, or in portions. The hearth must be hard enough so that no impression can be made with the finger. The concave hearth in the centre is six to eight inches deep and from six to ten feet in diameter.

The wall-ring is continued about two feet and a half above the hearth, and contracted for several inches on the top, on which the coppel rests. This ring has several apertures: one litharge hole, another hole opposite the fire-bridge through which the lead is introduced, and two holes for the tuyeres.

The hearth is charged either at once with so much lead as is intended for one trip, from five to twelve tons, or is filled with lead, and more added during the operation, in the same proportion, as litharge is drawn off. The latter way is preferable for this reason, that a smaller hearth can be used and less wood consumed on the same quantity of lead. A large hearth must be kept at a higher heat. This makes the hearth softer, and the mass absorbs more litharge. But the addition of lead, during cupellation, yields an impure litharge which, when reduced, renders a low quality of lead.

When the lead is charged, the coppel is placed on

the wall-ring, all clefts and little cracks are covered with loam and a slow fire started. By increased heat, the lead melts down, leaving all impurities on the surface. These impurities or dry scrapings (German, *Abzug*) are skimmed off. They contain a great deal of lead, copper, antimony, arsenic, also silver in oxydized and sulphuretted condition. If the pig lead is of pure quality, the scrapings are not regarded. The surface of the lead after the removal of the dry scrapings does not remain clear for a long time, but becomes again coated by a crust.

This crust, after the drawing of dry scrapings or after the melting of pure pig lead, must be scorified by increased heat and brought into fusion, admitting at the same time the blast. The fused crust or black-litharge scrapings, or froth (German, Abstrich), must be drawn off likewise. In the commencement it is spongy, black, has an imperfect metallic lustre, but assumes a gray greenish-yellow color at the end of the period. It contains metallic and oxydized lead, besides oxyds of zinc, iron, bismuth, antimony, arsenic, etc.; the antimony especially concentrates in it to a considerable amount.

Dry scrapings and the black litharge contain more silver than the litharge. The main part of the compound is oxyd of lead which has the property of taking up the sulphides of antimony and arsenic, by which again the sulphides of copper and silver are brought into the combination.

While the black litharge is drawn, heat and blast

must be increased, in order to hasten the scorification. To facilitate the drawing, some moistened coarse charcoal is thrown on the surface by which the froth assumes a spongy condition, when it is easily removed. The quantity of this froth depends on the quality of the lead.

When the black color of the froth changes into greenish-brown, assuming a more tough consistency, the oxyds of zinc, iron, and copper are mostly removed, and the black litharge consists in the greatest part only of the oxyds of lead and antimony. The more the latter is eliminated the more the yellow color of the oxyd of lead appears. The tough, scori-like condition disappears, and a short, scaly state appears. The mass flows no more down the floor, but hardens in front of the furnace. As soon as this change appears the drawing of froth is finished, and the temperature must be lowered.

The breast of the furnace or the litharge-bridge is cleaned, then a channel cut or scratched by an iron hook or saw-like instrument. Through this channel the litharge runs, wherein it is driven by the blast. The litharge channel, at a proper cupellation should be cut so deep only that the litharge ceases to flow, when the wind is stopped. The formation of litharge is difficult, if the antimony does not depart with the black litharge, because the lead has little inclination to oxydize before the antimony. This explains the long duration of some cupellations.

Generally the temperature must be kept as moderate

as possible, otherwise more lead and silver will volatilize. A hot litharge also eats the channel too much, and causes the escape of some lead. If the cupellation is kept too cold, the silver, which is in the unsufficiently liquid litharge, cannot come in contact well enough with the lead to be reduced again, consequently the litharge from too cold cupellation becomes richer in silver. At a too high temperature, more litharge is absorbed by the hearth, which is injurious, but this cannot be avoided entirely even by the best conducted heat.

On the edge of the bath, by escaping moisture and carbonic acid of the hearth a throwing up of bubbles is created, which follow the edge of the bath when the periphery of the lead decreases. The bubbles cease shortly before the brightening or concentration in the centre. It indicates too much heat if this boiling is violent, but it should not be too weak.

Care must be taken to keep always a sufficient quantity of litharge on the periphery of the lead-bath. The blast can be regulated in any required direction. In the first part of cupellation the blasts (of two tuyeres) have a divergent direction. According to the different position of the litharge the direction can be modified. At the end of the operation the blasts may cross each other. Hot air did not answer in all places.

The addition of lead commences as soon as a good run of litharge is observed. It is introduced by the feed-hole and placed on the brim of the hearth. Time and quantity of feeding depend on the progress of cupellation, and are indicated by the litharge and litharge-channel. At the end of the operation when less litharge is formed, the bath assumes a bright color. This is an indication that the period of brightening is approaching. The bath becomes covered with a net-like coat, movable on the convex surface, consisting of litharge-spots, between which the silver glances through. These spots grow larger, till at last the net breaks and the litharge slides to the sides, producing a peculiar shine which is called the brightening, distinguished by a play of colors. The colors are produced by the separation of the last of the oxyd of lead in very thin layers, through which the light passes, being reflected by the silver under a certain color. The kind of color depends on the thickness of the coating, which grows gradually towards the silver of the convex surface, producing the variety of colors, in a certain system. This is going on as long as oxyd of lead is emitted, and ceases when the silver becomes fine. The purification can continue under a strong heat, till to the required fineness, which in large furnaces however is not done on account of the consume of fuel by which the extensive space must be kept at a high temperature. The silver is generally taken out after the brightening, and the five or ten per cent. of foreign substances which are left in it, separated by refining.

After the brightening the wind is stopped, the silver cooled first with warm then with cold water and finally broken out by means of a long chisel.

The products of cupelling are the following:

a. The black litharge, or froth, is a very impure, ferriferous and copperous litharge, obtained in the first part of the operation. This litharge is not suitable for the market, neither for the reduction process. Generally it is melted with the ores. The black litharge contains sometimes as much lead as the yellow. The first litharge shows generally a brown or greenish color, caused by iron and copper. The affinity of copper and lead to the oxygen seems to be equal, as there is always some copper in the litharge from the beginning to the end of the process.

b. The poor litharge is the second litharge. It first looks yellow, but on getting cold, cracks in all directions. In the clefts arises a red, scaly, easily pulverizing product (the red litharge) while the rapidly-cooled crust retains its color and cohesion (the yellow litharge). The red litharge is the same chemical combination as the yellow, in an isomeric modification, differing only by structure and color. The production of the red litharge can be promoted by drawing the litharge from the cupelling furnace into hollow cylinders of sheet-iron, in which it cools, assuming a red color and the fine condition. This is the best article for the market. The yellow litharge is quite suitable for reduction.

c. The Rich Litharge.—This product is obtained in the last hours of cupellation. Being richer in silver this litharge is never sold but melted with the ore.

d. The Silver.—The fineness of the cupelled silver depends entirely on the time consumed after brightening.

e. The Hearth.—The hearth always retains more silver than the litharge, the silver of which is taken up by the lead during the longer contact. This is not the case with that litharge which draws into the hearth-mass. When the cupellation is finished and the furnace cooled down, the hearth is broken out and the heavy part, saturated with oxyd of lead, given over to the melting manipulation.

Cupellation in Movable Hearths.—This mode is practiced chiefly in England. These cupelling-furnaces differ from the German in two points: firstly, they have a fixed flat arch; secondly, the hearth-mass is generally bone-ash. The tests are prepared outside of the furnace in oval iron test-rings, four feet in the larger and two and a half feet in the smaller diameter. These tests are placed into the furnace from beneath, leveled, and the room between the test-ring and furnace-wall filled with fire-proof material. Generally, the English pig lead is so poor in silver that it must be concentrated by Pattinson's crystallization process.

Fine, pulverized bone-ashes are moistened with water containing some potash, then introduced into the test-ring and beaten in. The test is so cut out that the brim above is two inches, and at the bottom three inches thick, and the bottom itself one inch. On the front side the test brim is five inches wide, containing an opening for the litharge.

The test is brought into the furnace, dried, and then heated to red heat. The lead is introduced in fused condition, and the temperature raised till the crust melts, when the blast is thrown in, by which the litharge is driven to the front. By the addition of fused lead, the surface is always kept at the same level, till about five tons of lead are cupelled. The cupelling is then so far continued that the alloy may contain two or three hundred times as much lead as silver, whereupon the alloy is tapped, the tap-hole shut and the concentration continued in the same way. When so much rich lead is obtained that a silver-cake of from 3,000 to 5,000 ounces can be expected, the cupellation on it is executed in the described way.

Heated steam in place of wind yields a finer litharge.

CONCENTRATION OF SILVER IN POOR PIG LEAD,

By Pattinson's Crystallization Process.

SEC. 72. Most of the English pig lead is poor, being produced from galena, generally under twenty ounces per ton. The silver of such poor lead could not be extracted advantageously prior to the year 1833, when Lee Pattinson patented a new mode of concentrating the silver in argentiferous lead. This procedure permits the extraction if the ore contains only three ounces of silver per ton.

This method **is founded on the fact** that if an alloy **of lead** and silver **is melted in an** iron kettle and cooled, while often stirred, lead crystals will form at a certain temperature which are poorer in **silver** than the remaining liquid. The crystals are removed by means of perforated ladles, and several repetitions of **the same process** on the crystals and the enriched lead **will** yield a **very** poor lead of commerce and a rich part **for the** cupellation.

Besides the advantage of being enabled to extract small quantities of silver, this process yields also a purer and more **valuable** lead than that of reduced litharge. Although the cupellation of the enriched lead sustains **a loss of** at least five per cent. of lead, the average is nevertheless below two per cent., while the cupellation of the whole mass **would suffer** from seven to eight **per** cent. loss of lead. Very little or nothing is saved, comparatively, **in fuel and labor.** This process requires strong rather than skillful hands.

Lead, containing about five ounces per ton, offers the best advantage. At a higher amount of silver the expenses for labor and fuel increase, and the process is considerably delayed, still in many places lead with from forty to fifty ounces is concentrated with pecuniary advantage. Impure **lead,** containing antimony, arsenic, zinc, **or copper, is** not suitable for concentration, the separation of silver being imperfect. Such **lead** is first purified by continued heating in a reverberatory furnace or by stirring in a kettle with an unseasoned rod, and skimming off the impurities.

The good result of this process depends also on the right conduction of the temperature. If this be too low, the separation of crystals cannot take place, because the whole mass stiffens too soon. On the contrary, if too hot, no crystals can arise. Only a large quantity of lead will give a good result, because the transition of the liquid into the stiff state is only slow enough in a large quantity to give sufficient time for the removing of the crystallized lead. Many experiments have failed because operated with too small quantities.

Generally, in England, such lead is concentrated as contains from five to ten ounces of silver per ton, in lots of from 2·5 to five tons in four or nine cast-iron kettles, taking about fifty tons for one trip in succession. It requires several days to work up this quantity with two or four men.

At a battery of nine kettles, for instance, the lead with about ten ounces silver per ton is introduced in one of the middle kettles, and cooled down after it has been melted. The crust from the sides of the kettle is thrust into the liquid lead by means of an iron paddle, stirring at the same time the mass. With an iron perforated ladle (of which the handle is nine feet long, the ladle from twelve to fifteen inches diameter, six inches deep, with three-quarter inch holes in the bottom), the crystallized lead is dipped from the bottom of the kettle, and after the liquid lead drops off by shaking the ladle, it is laded into the next kettle on the left. This is done till about $\frac{4 \text{ or } 5}{14}$ of the original quantity, with an average amount of five ounces silver is transferred. The next $\frac{4 \text{ or } 5}{14}$ with about ten ounces silver, is laded temporarily into a flat kettle. The last $\frac{4 \text{ or } 5}{14}$ with twenty ounces of silver is transferred to the first kettle on the right. The lead from the flat kettle with ten ounces of silver, is introduced again into the first, with another portion of the original lead, and proceeded with in the same way as long as there is lead of ten ounces of silver at hand.

As soon as the kettle with the crystals on the left with five, and the kettle on the right side with the enriched lead containing twenty ounces of silver per ton, are filled, the separation is executed in both, and afterwards in all the following kettles, in the same manner as it was done in the first one.

From the last kettle on the left side the lead is not further concentrated, but laded into moulds for the market. It contains from one-fourth to one-half ounce of silver per ton. The enriched lead of the last pot has from two hundred to four hundred ounces of silver per ton.

The following scheme gives a view of the procedure of the work:

SCHEME OF THE PATTINSON PROCEDURE OF SILVER CONCENTRATION WITH NINE KETTLES.

O
BEGINNING POT WITH
10 ozs. Rich Lead
gives into
the

I
DESILVERING POT.
5 ozs. Rich Lead which gives
Crystals to II, and
Liquid back to O,

I
ENRICHING POT.
20 ozs. Rich Lead which gives
Crystals back to O, and
Liquid to II,

II
DESILVERING POT.
2½ ozs. Rich Lead which gives
Crystals to III, and
Liquid back to I,

II
ENRICHING POT.
40 ozs. Rich Lead which gives
Crystals back to I, and
Liquid to III,

III
DESILVERING POT.
1 to 5-4 ozs. Rich Lead which gives
Crystals to IV, and
Liquid back to II,

III
ENRICHING POT.
80 ozs. Rich Lead which gives
Crystals back to II, and
Liquid to IV,

IV
POOR LEAD POT.
¼ to ½ oz. Rich Lead
which is ladled into moulds.

IV
RICH LEAD POT.
180 to 250 ozs. Silver per ton
of Lead for Crystallization.

Extraction of Silver from Pig Lead with Zinc.

Sec. 73. The extraction of silver from argentiferous lead with zinc is executed in such a way that the molten lead is stirred with fused zinc several times, whereupon, on cooling, argentiferous zinc is eliminated on the surface of the bath, and may be skimmed with a perforated ladle, or it can be taken up in the shape of a disk. The zinc is then parted from the silver by acids (muriatic or sulphuric acid), or by distillation. The silver residue can be refined. The lead must be refined in a reverberatory furnace, on account of the zinc which it retains. The desilverization of lead, not too rich, can be accomplished very perfectly, but the separation of lead from zinc is imperfect. For this reason, and because this method depends on the price not only of the zinc, but also on that of chloride or sulphate of zinc, this method is not often used.

The more perfect the contact between lead and zinc is effected, the more complete the extraction of silver. Rich lead cannot be perfectly desilvered at once. A combination of Parke's and Pattinson's methods may answer in such a case. The first extracts the greatest part of the silver, and the poor lead may be subjected to the crystallization process, purifying it at the same time.

Comparative experiments of Parke's and Pattinson's methods have not yet been made sufficiently. According to Nevil, who tried both methods, Parke's procedure yields decidedly more silver. According to his statement, the loss of lead was one per cent., and of the zinc, three-fifths of the quantity of zinc alloyed with lead. In the establishment of Llanelly, where Pattinson's process was practiced heretofore in twenty pots, they have been replaced by two for the desilverization with zinc.

The quantity of zinc depends on the quantity of silver in the lead, and also on the amount of sulphur, arsenic, antimony, etc., which mostly join the zinc.

The small loss in metals, which occurs in this process, recommends it for poor lead which could not be desilvered otherwise with pecuniary advantage. At Carmarthenshire, in South Wales, this process is in use, and has been for several years. The operations are the following:

a. Melting of Argentiferous Lead with Zinc.—The lead is fused in a large iron kettle, holding about six tons. When this is done, some zinc is introduced in a smaller handled kettle, and heated by the ascending flame from the lead kettle. When the lead and zinc are fused, the zinc kettle is lifted sideways, and the contents poured in the liquid lead at once. Lead containing fourteen ounces of silver per ton, requires one per cent. of zinc. Four workmen stir the mass diligently four or five minutes, by means of iron curved rods, whereupon, the whole is allowed to rest five minutes. The eliminated scum is skimmed with large perforated ladles, made of sheet-iron.

b. Separation of Argentiferous Zinc from adhering Lead.— The plumbiferous zinc-scum is heated slightly in a fire-proof retort, not however above the melting-point of the lead, which liquates and collects in a basin, while the residue is drawn out at the rear of the retort.

c. Distillation of Argentiferous Zinc.—The argentiferous zinc is placed in pots, which contain three openings. One is the feed-hole, which is closed with a suitable brick during the retorting. On the bottom is another hole by which the residue is taken out. The third hole, on the side, serves for the escape of zinc-vapors. These retorts are placed over a grate and heated, whereby the zinc distills and is used over again for desilverization. The residue must be refined with a small addition of lead.

d. Cleaning of the Desilverized Lead.—In a reverberatory furnace with a low arch, the lead is heated rapidly with closed doors to a dark red heat, whereupon the air is allowed to enter the furnace through the door. The crust which arises on the surface of the lead is drawn off from time to time, and such temperature maintained as to promote the oxydation of zinc, but low enough to prevent the formation of much oxyd of lead. If the surface remains bright the lead is laded into moulds, and is a very pure article.

REFINING OF SILVER.

SEC. 74. The refining is a continuation of cupellation by which all impurities of the silver are oxydized and parted, but it is performed generally in a more concentrated space. The oxydation process is carried on in the cupelling furnace till the silver brightens; it contains several per cent. of impurities yet, as lead, copper, antimony, etc. A small quantity of lead, antimony, or arsenic makes the silver brittle, but copper does not injure its ductility. It may, therefore, be desirable under certain circumstances, to leave the copper in the alloy.

The refining is very simple, if the impurity consists only of lead; but if a considerable amount of copper, arsenic, and antimony, and less lead are alloyed with the silver, the separation of those substances by a mere oxydizing-melting takes a very long time to accomplish by the action of the atmospheric air. Besides this, the silver requires more heat if little or no lead is in it. In such cases, it is advantageous to add some lead; it being oxydized, eliminates also the other impurities.

According to Karsten, to one part of foreign substances, eighteen parts of lead should be used. In case the silver contains nickel or cobalt, an addition of some copper is advantageous. The formed litharge is not drawn out of the furnace, but is absorbed by the porous mass of the test (marl or bone-ashes, etc.).

The porous mass is either on a movable hearth, or a fixed one. The movable one is an iron bowl or an iron ring prepared with porous fire-proof material. These tests are placed in the refining furnace. The unmovable test does not differ much from a cupelling furnace except in smaller dimensions, which are relative. The oxydation is effected either by blast or draft. The refined silver is not **chemically pure.**

The sign of **pure silver is the** so-called "spitting" of the silver. **When** the surface **of the silver** begins to harden **an eruption** of metal, assuming different forms, takes **place;** and during **this process** small **globules** are often thrown off. This phenomena had **been attributed to a** physical cause, but it was discovered by **Lucas that** silver possesses **the** remarkable property of absorbing, while in melted **condition, a large quantity of** oxygen gas (at least twenty-fold volume **of the** metal) which absconds rapidly **during the** refrigeration of the metal, forming figures one or two inches high, whereby some silver is often scattered.

By a very slow cooling **the** spitting **can** be avoided. The oxygen can be ascertained very easily **by** throwing charcoal-dust on the place where the spitting occurs. A very lively burning **of the** carbon will be perceived. The **silver** absorbs oxygen from saltpetre **if** covered with it when molten, but when the contact with **the air** is interrupted by a cover like **common salt,** no spitting will occur.

A small amount of gold does not impede the spitting, but it does not appear if some lead **or copper** is alloyed with silver.

After the refining, the test-mass or hearth is found **to contain, besides some** metallic and oxydized silver, also **the** oxyds of foreign metals, which were alloyed **with** the silver. The silver in such hearths may sometimes amount to **six** hundred ounces per ton, for that reason it is **always given** over to the smelting process.

The usual methods of refining are the following:

A. Refining on movable **hearths or tests.**

1. Before the blast.

2. Under the muffle.

3. In **reverberatory** furnaces.

B. **Refining** on fixed hearths in **reve**rberatory or **draft furnaces.**

Refining in small **draft** furnaces is the most advantageous and simplest method. **The** least advisable is the refining under the muffle, which consumes the most fuel, but in regard to cleanness and proper work **is** preferable. The refining before the blast is more difficult.

A. Refining on Movable Tests.

Refining before the Blast.—In regard to fuel, this mode **is** economical and very **suitable** for impure silver, which requires a powerful oxydation, especially if some copper is allowed to remain in the silver. It needs some exercise **to** conduct the process, so **as** to keep the silver always at **a proper** heat. Using the blast, the silver is disposed **to volatilize.**

In Freiberg, this procedure is practiced in such a manner that the silver in portions of from three hundred and twenty to four hundred and eighty ounces is melted before the blast in iron tests, which are lined by stamped marl. These tests are from nine to eleven inches wide, and about three inches deep. When the silver is introduced, some live coals are placed before the tuyere, and around the test there is a ring of sheet-iron filled with charcoal, then the blast is thrown in. The blast yields about eighteen cubic feet of air per minute. When the silver is molten, the ring must be removed, the glowing coal drawn from the surface of the silver, and the heat maintained by placing some thin pieces of wood between the silver and blast, and covering the wood with some live coal, so that only the flame is carried over the surface of the silver. Care must be taken not to let the silver refrigerate, and also to avoid the contact of coal and metal. The process is continued under frequent stirring, till the litharge spots cease to appear. A small iron hook is then dipped into the silver and the adhering drop examined. It is sufficiently fine, if the dark red-hot, pear-shaped drop, appears entirely free of spots, spitting on cooling, and assuming at the same time a perfect white color. The metal in the test is perfectly lustrous. The blast is stopped, the silver cooled cautiously with water, then taken out and cleaned. One trip takes from one to one and a half hours, consuming 2·2 cubic feet of charcoal, and 1·1 cubic feet of wood.

Refining under the Muffle.—This method consumes more fuel than the preceding, but it is cleaner, easier, and more correct work. This mode is more suitable for a poorer sort of silver.

The tests are of about the same size and prepared in the same way as the preceding and placed in a corresponding hollow place of the furnace, as level and secure as possible. The silver is then introduced, and the test covered with the muffle. The whole arrangement is in a small, vertical furnace, widened in the middle, with an aperture in the front. On the periphery at the bottom, are six little draft-holes, each half an inch square. By these holes the draft is regulated. The draft-holes of the muffle must be carefully covered with pieces of broken muffles, then the charcoal in large pieces introduced around the muffle. The aperture is closed by a tile with a hole in it, through which the inside of the muffle can be seen. The furnace is filled with charcoal and the fire started.

When molten, the silver must be stirred often by means of an iron hook, and the process continued by closed or open muffle, as the temperature may require, till no forming litharge is perceptible, and the metal looks perfectly bright. When this state is obtained, the silver-cake is cooled with water. The metal will soon commence to spit. An iron hook is pressed into the silver, on the spot where the spitting takes place, and by means of it, the silver cake is taken out. At Oker on the Hartz, eight hundred ounces of the introduced silver yields seven hundred and sixty ounces fine silver in five hours.

Procedure of the Mint at Clausthal.—The iron test-bowl is filled with wood ashes, well lixiviated, and beaten in level with the edge, whereupon a cavity is cut out about twelve inches in diameter, and from three to four inches deep. There are three furnaces of the preceding description, each for one muffle, and two larger ones, in each of which four muffles are placed. The latter furnaces are provided with four draft-holes on the back wall. The smaller ones have holes on two sides, and one behind. The tests are placed in the furnace as above described, charged with seven hundred ounces, and by closed muffle, melted down in about two hours. After this the muffle is opened and the metal stirred for some time with an iron hook, then shut up for half an hour, and then stirred again. This is repeated three times at intervals of half hours, whereupon the last heat is given, half an hour long. During the stirring, litharge eyes are produced on the surface of the bath. At the end of the operation they disappear. As soon as the silver looks perfectly bright, it is cooled by pouring water on it. The spitting is impeded by keeping an open hole in the centre of the cake with an iron hook.

B. Refining in Reverberatory Furnaces.

The reverberatory, or flame furnaces, have movable or fixed hearths, and are covered with an arch or movable cap. They do not differ much from the cupelling furnaces, save the size. The heat is produced by wood, coal, or gas.

The **test-rings,** or brick hearths, are prepared either with **lixiviated wood** ashes, or with fine, sifted marl. **The properly cut and** dried test is placed in the furnace, the silver introduced, covered with small charcoal, **and** heated to nearly white **heat at** well closed doors. To avoid losses of silver, **it is a** principal rule to cover it with fine charcoal **or saw-dust, and to** melt it down quickly, and to **refine it at a** lower temperature. The silver **volatilizes at the beginning of the** white heat, especially under draft; this must be kept from the silver by the cover of charcoal. The more antimony, **arsenic, or lead, the silver** contains, the easier it volatilizes, requiring **the charcoal cover** to be kept on it so **much** longer. **When molten, and** no more silver is added, the slag is **drawn off,** the **metal** stirred, and the surface cleaned. **The** temperature is so conducted, that the **eyes, swimming on the surface, should** not glide too fast **towards the** sides. At **a** too high temperature, the silver assumes a vibrating **motion, many** pearls hasten from the middle **towards the** periphery, and the silver does not adhere to an iron **rod,** if quickly dipped in and withdrawn. The temperature is **too** low, when on the **edge** of the silver bath **commencing refrigeration** be observed.

If the fusion is not easily obtained, some lead, from four to **six per cent.,** may be added. Easy refrigerating and slow refining silver requires some copper. If, after some stirring, **a** bright surface is observed, the assay **must** be taken, **by** dipping **a curved** iron rod about **half**

an inch deep into the silver till the latter assumes a pear-like shape. It is pure if no spots of litharge or oxyd of copper are perceived on it, and if the silver runs off, or endeavors to drop off, when dipped again. In the reverse case, the refining must continue. When fine, the silver is cooled with water and removed. In Freiberg, the refining furnace is charged at once with from twelve hundred to thirteen hundred pounds of silver.

Refining in Crucibles.

Generally only the purer silver is subjected to refining in crucibles. The melting is performed either in a black-lead or cast-iron crucible, either with charcoal or with the flame of coal, in reverberatory furnaces, in which the crucible is placed. The black-lead crucibles at Freiberg have been replaced by cast-iron ones in flame-furnaces, offering many advantages over the former in draft furnaces, with charcoal.

The loss of silver by volatilization was brought down to the minimum, while, when exposed to the strong draft of the crucible furnace in a black-lead crucible, it was considerable. The use of coal proved more economical than charcoal, and the iron crucibles stood longer than the black-lead crucibles. At a white heat in an uncovered crucible the silver loses about one per cent. per hour.

At Przibram (Bohemia) the crucible is charged by degrees, with from 4,800 to 9,600 ounces of silver, and melted in five or six hours. A mixture of two parts lixiviated ashes, and one part bone-ashes, is introduced on the molten silver, by means of an iron ladle, stirred in such a way on two places, as to obtain two openings in the mass, through which the oxygen of the air can come into contact with the silver. The oxyd of lead is taken up by the porous mass, and this removed by means of a ladle. This operation must be repeated at gradually longer intervals, till the silver commences to boil and to show a bright surface. After this, a mixture of one ounce and a half of borax with the same quantity of saltpetre, is introduced, and then the slag skimmed off, after a quarter of an hour's time. The silver is immediately covered with charcoal dust and a good heat applied for a quarter of an hour, when the silver is dipped into moulds. This method suffers a small loss in silver, consuming comparatively very little fuel, but the operation requires more time.

CHAPTER IV.

EXTRACTION OF SILVER IN THE WET WAY.

A. EXTRACTION BY QUICKSILVER OR AMALGAMATION.

SEC. 75. Where gold and silver ores were found in large quantities, but no fuel, it was necessary to search for other means of reduction than heat, and this was found in quicksilver. Poor silver ores, containing small quantities of lead, are also suitable for amalgamation.

The extraction of silver by quicksilver is founded on the property of the latter to form an alloy with the silver, which can be separated by heat. A procedure for extracting the silver from ores by quicksilver was first published by Bartolomé de Medina in Mexico, in the middle of the sixteenth century. In the year 1784 the amalgamation was introduced in Europe first by Born, at Vienna, in copper kettles, then by Gellert in tubs, and finally by Ruprecht, in barrels.

The amalgamation, compared with melting, has the advantage of saving fuel, rendering the silver in a short time,—unlike the melting by which the silver is carried through many intermediate products, causing thus more loss in silver, expense of time and money. There are chiefly three modes of amalgamation:

1. European Barrel-Amalgamation.

2. American Amalgamation.

   a. Heap-Amalgamation (Patio).

   b. Kettle-Amalgamation.

3. Combined European and American method.

European Barrel Amalgamation.

The European barrel-amalgamation is more perfect than the patio-amalgamation, yielding the silver in much shorter time and with less quicksilver consume, but makes the use of more machinery and fuel necessary. In both methods the silver must be converted into a chloride and decomposed in the barrels by iron, in the patio by quicksilver. The European method is applied to ores, matts, and black copper.

Amalgamation of Silver ores.

All silver ores are not suitable for amalgamation. Copper, lead, antimony, arsenic, and zinc, are not agreeable customers, partly because they enter the amalgam, and because they effect losses of silver and quicksilver by volatilization, also impede the amalgamation and produce richer tailings (the last is caused by lead). The zincblend needs a strong heat, but decomposes the salt very little. Argentiferous blend loses a great deal of silver by volatilization.

Iron acts favorably; manganese, nickel, and cobalt are not injurious. Experience shows that, concerning earthy matters, quartzose ores yield more silver but less pure amalgam, and cause larger loss in quicksilver, while calcareous ores work better on quicksilver, giving also purer amalgam, but less silver. The best economical result is obtained by mixing both kinds. Clayish ores are amalgamated with difficulty.

The ores best suitable are the pyritous, sulphuretted silver ores, without regard to richness, but salt and quicksilver must be regulated according to the richness. In roasting the ore with salt, a certain amount of pyrites is necessary for the formation of chlorine, which is ascertained by an assay for matt. If there is no pyrites in the ore, iron or magnetic pyrites, matt or green vitriol, must be added.

The Roasting—With the intention of forming chloride of silver—is one of the most important preparations of the ore. It must be executed with great care. In Freiberg, one charge used to be four hundred and fifty pounds. It is introduced into the dark red furnace, carefully spread on the hearth, stirred and the lumps broken with a long-handled hammar. After this, a moderate fire is started and continued, gradually raising the heat for about two hours. At this time, the furnace having assumed a light red heat, the sulphur begins to burn. In the first period, white vapors arise chiefly of water, antimony, arsenic, and zinc.

With the beginning of the burning of sulphur or desulphurization, the second period commences. Under constant stirring the fire must be made to go down, as the burning sulphur creates a sufficient temperature. In this period, lasting again two hours, the principal action on the ore is effected in an oxydizing manner by the atmospheric air. Sulphurous acid, basic, and neutral sulphates are formed, as well as free oxyds; but some sulphides remain also undecomposed, resulting from higher sulphur combinations.

When the odor of sulphurous acid has disappeared, and the temperature has sunk to dark red heat, the last period begins. The heat must now be raised, under continual stirring; the ore swells up, enlarging its volume, greenish-gray vapors are emitted, and the last period is generally finished in three-quarters of an hour, and the ore drawn out while the evolution of gases is still going on.

The sulphates, especially the sulphate of iron, act in this period on the salt, whereby partly free chlorine and partly hydrochloric gas is formed, the latter in the presence of water-vapors which are always supplied by the fuel and air. In passing through the ore, the chlorine decomposes principally the sulphides which escaped decomposition by heat, forming volatile chlorides of sulphur, arsenic, antimony, iron, and zinc, also unvolatile chlorides of gold, silver, copper, lead, nickel, cobalt, iron, and manganese. The hydrochloric gas changes principally the oxydated metal combinations into chlorides. The chlorination of silver already begins in the second period of roasting, but principally in the third. If the roasting was well performed, all silver must be changed into a chloride, otherwise some sulphurets of silver remain undecomposed and the tailings become rich.

Examining the ore during the roasting for chloride of silver, the following procedure will show the result:

A sample is taken from the furnace and one ounce of it weighed out, introduced into a filter and lixiviated with a hot solution of salt until a clean piece of copper does not show a white coating of precipitated silver. The ore is then dried and assayed for silver.

The roasted ore must be sifted and ground. The coarse part is ground separately and roasted over with two per cent. of salt; the fine-sifted and ground ore is given over to the amalgamation.

Amalgamation.—This is performed in barrels made of oak or pine wood. The first period is the preparation of the pulp. For this purpose water (three hundred pounds) is led into each barrel and 1,000 pounds of roasted ore introduced, also one hundred pounds of wrought-iron. The barrels are shut and started with a speed of fourteen to sixteen revolutions per minute. The water dissolves all soluble salts and exposes in this way the particles of chloride of silver. In this period the iron acts decomposingly on the chlorides of iron and copper, changing them into sub-chlorides which are not injurious to the quicksilver. The chlorides of copper and iron would transform a part of the quicksilver into sub-chloride which would be lost in the tailings. The dissolved chlorides of silver, gold, and copper are reduced by the iron to a metallic state. After two hours' preparatory run, during which the pulp has obtained the proper consistency, the barrels must be arrested and the second period (the amalgamation) commences. To each barrel is now added five hundred pounds of quicksilver. When secured, the barrels are started again for twenty hours with a speed of twenty to twenty-two revolutions per minute. By a galvanic action the further decomposition of the chloride of silver takes place. The galvanic action is created by the positive iron, the negative quicksilver, and the dissolved salts as conductor. The negative chlorine combines with the iron, the positive silver with the negative quicksilver. Temperature is produced by these actions. Besides the chloride of silver, also other metal combinations are decomposed by the galvanic action, especially copper, lead, antimony, and gold, which combine with the quicksilver.

The third period is the separation of amalgam and residue. The amalgam, disseminated in the pulp is separated by water, with which the arrested barrels are charged over two-thirds full, and again brought in motion for two hours, revolving eight or nine times per minute, whereupon the amalgam and quicksilver is discharged, first, **through a** small **hole** in the plug, then the residue **through a** large **hole** (five to six inches in diameter).

The residue of each five barrels **runs** into one agitator, in which the amalgam still left **settles to the bottom,** **while** the tailings are discharged gradually through **three** holes, one above the other.

The amalgam, when discharged from the barrel **is led** directly into canvass filters. **The** quicksilver, pressed by its own weight, **runs through the** cloth into a reservoir.

If it ceases **to run,** the separation of the remainder of the quicksilver **is** performed **by** pressing with **the hands** or press.

Amalgamation of Copber-Matt.

The copper-matt amalgamation makes the same manipulation necessary as the working **of ores.** The matt, however, after a first **raw** roasting is drawn out, mixed with salt and lime, moistened, and the mass after **proper** drying, roasted again. This modification is **based on** the amount of copper in the matt, **with the object of** changing **the chloride of copper into oxyd of copper,** because **the chloride of copper** chloridizes a part of the quicksilver **when in contact.** In the first roasting of the matt, sulphates of copper, silver, and iron are formed. **By the action of** salt, they are transformed into the chlorides of copper, and silver, **and sequi-chloride of iron.** The lime decomposes the chloride **of**

copper and sesqui-chloride of iron into hydrated oxyds, without changing the chloride of silver. Herein the experience that an addition of lime diminishes the loss of quicksilver is accounted for. The subsequent or the second roasting is performed at a higher temperature, effecting the perfect chlorination of the silver.

The amalgamation of the matt is advantageous, if there is no gold in the ore, no lead ore to be had or the fuel is scarce, otherwise melting with lead would be preferable.

Amalgamation of Argentiferous Arsenides (Speiss) obtained in Melting.

The arsenides, obtained in melting of cobalt ores, are analogous to the matts. In the first place we have arsenic, in the second sulphur combined with metals. The amalgamation of Speiss is executed in Saxony. It requires a very careful roasting and fine grinding, and an addition of sufficient salt. As there is not sufficient sulphur in the ore, by which the salt can be decomposed, two and a half or three per cent. of green vitriol must be added for that purpose. The tailings contain five ounces of silver to the ton, but the average loss is about thirteen per cent.

Amalgamation of Argentiferous Black Copper.

Experience proves that the alloy of silver and copper, without sulphur, decomposes the salt forming chlorides of silver and copper. The closer the contact between salt and copper, the more chloride of silver is formed. The chloride of copper transfers also chlorine to the silver, being thus reduced to sub-chloride.

Black copper requires, as well as the ores, an abundance of salt, and deserves preference to the copper-matt amalgamation on account of its simplicity and more perfect extraction of silver. This method has a decided advantage over the liquation process, on black copper, free from lead and gold.

In Schmölnitz (Hungary) the black copper contains from one hundred and ten to one hundred and fifty ounces of silver per ton, and eighty-five–eighty-nine per cent. of copper. This copper is heated red in a reverberatory furnace and stamped in glowing condition, then sifted and ground to fine powder. For the roasting, four hundred pounds is taken to each charge and oxyd with seven to nine per cent. of common salt. After roasting, it is ground again, and then amalgamated. The barrels are charged with from 1,200 to 1,500 pounds each, one hundred pounds of copper balls, four hundred pounds of quicksilver, and the required quantity of water. The loss of silver is four and three-fourths per cent., of which only two and a half per cent. remains in the residue. The pure copper resulting from this process contains about ten ounces of silver to the ton. At Tajova, a loss of twenty-seven per cent. of silver occurs and the copper retains twenty ounces per ton.

At Offenbanya (Transilvania) the black copper contains some lead, and on this account three per cent. of saltpetre is added with the salt in roasting.

B. EXTRACTION OF SILVER BY DISSOLUTION AND PRECIPITATION.

Augustin's Extraction with Solution of Salt.

SEC. 76. This method is based on the solubility of the chloride of silver in concentrated solution of common salt, from which it is precipitated by copper. If water is added to a salt solution with dissolved chloride of silver, the latter will be again separated in form of chloride of silver as a white precipitate. This fact, on which Augustin based his extracting method, has been known a long time. It was first executed on coppermatt at Gottes-belohnung works, and since then at other places, also on other argentiferous products and ores. The best results have been obtained on copper-matts.

Extraction of Silver from Copper-Matt.

The Copper-Matt—Suitable for extraction must be as rich in copper as possible, otherwise the residue becomes too rich in silver. Poorer copper-matts are therefore concentrated by melting in reverberatory furnaces, being thus enriched to from sixty to seventy per cent. The matt must be free from or should contain very little lead, antimony, zinc, or arsenic, or else a considerable loss in silver might occur.

Lead—Makes the roasting troublesome, having an inclination to melt, which counteracts the formation of the chloride of silver. At the same time chloride of lead is created, which would also be dissolved in the salt solution, causing an impure precipitation of silver. Sulphuret of zinc, to a certain amount, is injurious in a chloridizing roasting, because, when changed into a sulphate, it requires a high temperature to drive the sulphuric acid out, which must be done, otherwise too much chloride of zinc will be formed. But at a high temperature the matt softens by commencing to melt, causing thus richer tailings and also a larger loss of silver by volatilization. For this reason, zinciferous matt must be freed from zinc by concentration-smelting in reverberatory furnaces.

Antimony and *Arsenic*, in a chloridizing roasting, produce also antimonate and arsenate of silver, neither of which can be decomposed by the chlorine. Introduced steam, however, forms hydrochloric acid, by which the above combinations are decomposed, resulting in chloride of silver and volatile chlorides of antimony and arsenic.

The fine pulverized matt, principally consisting of sulphides of iron, copper, and silver, is first roasted without salt, whereby the iron is converted into sulphate of iron, then sulphate of copper is formed, and finally sulphate of silver. The temperature, under which the sulphate of silver is produced, is so high that almost all the sulphate of iron and a great part of the sulphate of copper will be decomposed, so that at the end of roasting the mass consists chiefly of the oxyds of iron and copper, some sulphate of copper and sulphate of silver, as well as some undecomposed sulphides. From time to time samples are taken, to ascertain the condition of the roasting. The sample is placed on a filtering paper and lixiviated with water. The blue color, according to its intensity, shows the presence of more or less sulphate of copper. An addition of salt solution gives a white precipitate of chloride of silver.

In the second roasting period the oxydation is continued for a while at an increased heat, in order to decompose all sulphides of iron, copper, and silver. When a sample, taken from the furnace gives a feeble blue-colored water, and with salt solution a strong precipitate of silver, the purpose of the oxydizing roasting is accomplished. Now is the time to introduce the salt and to change the roasting into a chloridizing one.

The sulphuric acid, which is principally combined with silver and to some extent with copper, expells the chlorine of the salt, forming sulphate of soda and sulphurous acid. The chlorine, penetrating the mass, creates chloride of silver; but as there is always a superabundance of chlorine, it creates also some chloride of copper and iron. In this condition the roasted matt is subjected to lixiviation with hot concentrated solution of salt. The desilverization is promoted and perfected if the solution acts under pressure upon the matt. From the solution, the silver is precipitated by metallic copper. The copper precipitating the silver, enters into solution and is precipitated by the iron. The subchloride of copper is produced by decomposition of chloride of silver, and also by decomposition of the chloride of copper in the solution, with metallic copper.

The products of this process are:

1. Cement-silver.

2. Cement-copper.

3. Residue melted for copper.

4. Green vitriol or sulphate of iron, and,

5. Glauber's salts.

In Freiberg this method offered also a great advantage over the liquation process, and is executed in the following way:

The matt, after stamping and sifting, is subjected to roasting in portions of from three hundred to four hundred pounds. The roasting is performed in a double furnace, one hearth above the other. Each charge remains four hours on the upper and four hours on the lower hearth. The roasted matt is then ground fine, and roasted over on the lower hearth for one hour with an addition of five per cent. of salt.

The roasted matt in now subjected to the lixiviation. For this purpose it is elevated to the upper story, where the lixiviating tubs are arranged. They are three feet nine inches high, two feet eight inches in the upper and two feet four inches in the lower diameter, and are filled with from four to six hundred pounds of matt. Each of these tubs has a filtering apparatus on the bottom; first a wooden cross then a perforated wooden bottom, then a layer of straw and over this a piece of cloth, made tight against the staves of the tub, by being stretched upon a hoop. On the top of the tub, when filled with matt, there is placed a perforated wooden disk. The concentrated hot salt solution is conveyed in a trough to the tubs, into which it drops, coming out through the cock at the bottom saturated with the chloride of silver. This cock is first opened for the escape of vapor, then shut for a quarter of an hour and opened again, taking care that as much salt solution flows into the tub as comes out through the cock.

The lixiviation is performed in three periods. The first, with concentrated solution, takes ten hours. The second, also with concentrated solution, continues till a piece of clean copper does not become coated white; and this may require ten or twelve hours longer. The last lixiviation is done with clear water. The products obtained are:

1. *Residue.*—The tubs are transported on a railtrack to a discharge place, an assay is taken, and in case the tailings should prove to be unsufficiently desilvered,

they must be roasted over again with salt. They contain from forty to sixty-five per cent. of copper.

2. *Silver Salt Solution*—Which is conveyed into precipitation tubs.

3. *Lixiviating Water.*—This water, obtained from the third period, contains salt. It is led into a basin and used again for lixiviation.

The cement-copper, obtained by precipitation with iron, is introduced into the precipitating tubs, about six inches deep, in which the silver-containing solution is conveyed. The tub is provided with a filtrating apparatus, through which the desilvered solution is carried into other tubs where the copper is precipitated by iron and used again for precipitating the silver.

The precipitated silver is removed at intervals of eight days from the copper on which it accumulates. Sub-chloride of copper and chloride of lead are separated from the precipitated silver by washing it with water and diluted muriatic acid, then melted and refined in crucibles.

C. ZIERVOGEL'S EXTRACTION OF SILVER WITH WARM WATER.

SEC. 77. Ziervogel's method of silver extraction from matts is more simple and cheaper than the preceding. No salt is used here.

In roasting pure argentiferous copper-matt sulphate of iron is first formed, then sulphate of copper, and finally sulphate of silver. The formation of the last sulphate requires such high temperature that it decomposes again the first two sulphates, created at a lower heat. The sulphates of iron and copper are decomposed into oxyds, and sulphuric acid which escapes, while the sulphate of silver remains undecomposed. This sulphate is soluble in hot water, and can be precipitated by copper.

It has been observed generally that Ziervogel's method is simpler and cheaper, and the lixiviation is effected quicker, but the roasting process is a great deal more difficult, purer matts are necessary, and the tailings are richer than from Augustin's process.

The matt is pulverized to a fine powder, and, in charges of five hundred and seventy-five pounds, roasted. There are double furnaces used for this roasting. The upper floor is charged with the matt, and this stirred for one hour and a quarter; the lumps, if any, are mashed and the mass changed so that the cooler part is brought nearer to the fire, and *vice versa*. The stirring is continued for one hour and a half, the mass changed again, and now from twenty to twenty-five pounds of coal-dust are added, well mixed, for ten minutes and then conveyed to the lower hearth, through a flue which was covered with an iron plate. The matt is now spread on the red-hot hearth, and stirred for about one hour and a quarter without using any fuel. The mass will first ignite on account of the coal dust. It is then raked together and the different heated parts change places. A sample is taken from the bridge-side with an iron ladle, emptied into a porcelain cup, so as to divide it in two partitions. Cold water is then slowly introduced on one side till it penetrates the sample and comes forth on the other. The water must appear feebly-blue colored, and several grains of salt must not produce a darker hue but precipitate the silver as a white substance.

If such reaction is observed the matt must be shifted, firing and stirring continues; another sample is taken at the bridge-side and examined in the same way till a corresponding result is obtained. But if the water should show a green color from sulphate of iron, or a deep blue from sulphate of copper, the stirring must be continued.

As soon as the roasting is finished which is indicated by the above sample, a sign is given to the upper workman to add coal-dust to the matt of the upper hearth. The roasted matt from the lower floor is removed and sifted through a sieve of sixteen holes to the square inch. The coarse part of the sifting is added to each charge in quantities of twenty to twenty-five pounds and reroasted.

The fine-sifted matt is brought to the cooling place, where it remains for six or eight hours. The result of this process depends so much on the roasting that at Freiberg an arrangement is made, according to which the roasters receive a premium, if the tailings are found poorer than the allowance.

The lixivation is carried on with hot water in wooden tubs similar to those described in Augustin's process. They are arranged in galleries. On the first there are eight wooden cylindrical lixiviating tubs, provided with a filtering apparatus. The tubs are two and a half feet high and the same width. At the bottom there is a cock through which the filtrate is conveyed into clearing tanks, thirty feet long, one and one-half feet wide, and one and one-half feet deep, divided into two partitions. The lye enters the first partition, then over the parting board into the second. From thence it flows through ten cocks into as many precipitating tubs, which are posted on the second gallery. These tubs are also provided with filters and cocks. Each of them contains ten pounds of granulated copper, and over it in each two hundred and fifty pounds of black copper, fourteen inches long, five wide and one inch thick.

The fluid, after passing the tubs of the second gallery, flows into precipitating tubs on the third gallery of which there are five in number, containing likewise copper for precipitation. From this last gallery the fluid flows into a leaden basin, wherefrom it is pumped into another leaden basin and heated by steam till it assumes a temperature of 158° Fahrenheit.

The fine-sifted matt, after having rested for about eight hours is cooled down to about 158° Fahrenheit. In this state it is introduced into the precipitating tubs of the first gallery (four hundred pounds in each) and water added till it commences to run out at the cock on the bottom. The water is previously heated by steam to 158° Fahrenheit.

The water must be stopped as soon as it begins to run through the cock, and then so much of the heated fluid of the leaden basin is allowed to flow on the matt, till a sample from the outflow does not show a precipitate of chloride of silver by salt solution. This lixiviation takes two hours and a half. The tailings are removed and given over to the copper-smelting process. The precipitation of silver by copper in the precipitating tubs, is similar to the operation described in Augustin's process.

D. PATERA'S SILVER EXTRACTING PROCESS.

SEC. 78. By a chloridizing roasting the silver is converted into chloride of silver, then dissolved in hyposulphite of soda, and precipitated by polysulphide of sodium.

Cold, diluted hyposulphite of soda dissolves the chloride of silver sooner than a hot concentrated solution of salt. The polysulphide of sodium precipitates the silver as a sulphide, the reduction of which is performed by simply calcining under a muffle.

The ore is first subjected to an oxydizing roasting, in which the heat must be increased slowly by degrees, especially if the ore contains many different sulphurets.

When red heat is obtained, some steam is introduced in the furnace and as much as possible without diminishing the necessary heat. The ore must be constantly stirred. In about four hours the first roasting is completed and the ore is discharged. This must be ground fine, mixed with from six to ten per cent. of salt, or more according to the richness, and subjected to the second—chlorodizing roasting. Two or three per cent. of calcined green vitriol is added for the decomposition of salt. When red-hot, the steam is used again, and diligent stirring kept up for five or six hours. Rich ore may require a longer roasting by several hours.

The roasted ore is now ready for lixiviation. There are chlorides of different base metals in the ore, which are soluble in water, while the chloride of silver resists, for this reason two lixiviations are adopted: the first with hot water, which will dissolve and carry off the chlorides and sulphates of copper, iron, zinc, cobalt, etc., and the second with hyposulphite of soda which dissolves the chloride of silver. If the roasting is not well performed so that some sulphate of silver remains in the ore, this will be also dissolved by the water and carried out with the base metals.

Charges of four hundred pounds of roasted ore are thrown into tubs, prepared for lixiviation like those used and described in Augustin's process. Hot water is conveyed to the tubs for about six hours by a constant stream, purifying thus the mass of base metals. After this cold water is used for the purpose of cooling the ore, which is not allowed to be warm for the second lixiviation.

When cold, the ore must be removed in other tubs of the same description, but smaller. A cold solution of hyposulphite of soda is brought into the tubs, and continued till in a taken sample no precipitate of silver is obtained by a solution of polysulphide of sodium.

The solution, now in precipitating tubs, is mixed together under good stirring with the precipitant (polysulphide of sodium). The precipitate (sulphide of silver) appears black. Attention is necessary to use the right proportion of the precipitant, in order to obtain a neutral liquor after all the silver is precipitated. This fluid is used over again for lixiviation, but a surplus of the precipitant would render the fluid unsuitable for that purpose and precipitate some silver in the mass during the lixiviation. If no more precipitation is observed, a small quantity of the solution is taken in a glass tube and some drops of polysulphide of sodium added. If a black precipitate is produced in a slight degree, some more of the precipitant must be added to the solution in the precipitating tubs; but if, on the contrary, no precipitate was obtained in the glass tube, another sample must be taken, and tried with some drops of the original solution. If then a black precipitate is formed, too much of the precipitant was introduced into the precipitating tub and must be neutralized by addition of the original solution.

The precipitated black sulphide of silver deposits in

bags which are well pressed, dried, then washed with warm water in filters, dried again and heated under muffles, to which air has access. The sulphur burns off and the silver remains in metallic condition. This is melted finally in crucibles with an addition of some iron and borax.

TABLE I

SHOWING THE AMOUNT OF SILVER OR GOLD IN OUNCES CONTAINED IN ONE TON OF ORE (TWO THOUSAND POUNDS) FROM THE WEIGHT OF FINE METAL OBTAINED IN AN ASSAY OF HALF AN OUNCE OR TWO HUNDRED AND FORTY GRAINS OF ORE.

Ounces of Fine Metal per Ton of Ore of 2,000 Pounds.

Fine Metal — Thousands of the unit of 10 grains.	One ton of ore will contain — Ounces.	Fine Metal — Thousands of the unit of 10 grains.	One ton of ore will contain — Ounces.	Fine Metal — Thousands of the unit of 10 grains.	One ton of ore will contain — Ounces.
		·030	36·45	·060	72·90
·001	1·21	1	37·66	1	74·11
2	2·43	2	38·88	2	75·33
3	3·64	3	40·09	3	76·54
4	4·86	4	41·31	4	77·76
5	6·07	5	42·52	5	78·97
6	7·29	6	43·74	6	80·19
7	8·50	7	44·95	7	81·40
8	9·72	8	46·17	8	82·62
9	10·93	9	47·38	9	83·83
·010	12·15	·040	48·60	·070	85·05
1	13·36	1	49·81	1	86·26
2	14·58	2	51·03	2	87·48
3	15·79	3	52·24	3	88·69
4	17·01	4	53·46	4	89·91
5	18·22	5	54·67	5	91·12
6	19·43	6	55·89	6	92·34
7	20·65	7	57·10	7	93·55
8	21·86	8	58·32	8	94·77
9	23·08	9	59·53	9	95·98
·020	24·30	·050	60·75	·080	97·20
1	25·51	1	61·96	1	98·41
2	26·73	2	63·18	2	99·63
3	27·94	3	64·39	3	100·84
4	29·16	4	65·61	4	102·06
5	30·37	5	66·82	5	103·27
6	31·59	6	68·04	6	104·49
7	32·80	7	69·25	7	105·70
8	34·02	8	70·47	8	106·92
9	35·23	9	71·68	9	108·13

Ounces of Fine Metal per Ton of Ore of 2,000 Pounds.

Fine Metal—Thousands of the unit of 10 grains.	One ton of ore will contain—Ounces.	Fine Metal—Thousands of the unit of 10 grains.	One ton of ore will contain—Ounces.	Fine Metal—Thousands of the unit of 10 grains.	One ton of ore will contain—Ounces.
·090	109·35	·120	145·80	·150	182·25
1	110·56	1	147·01	1	183·46
2	111·78	2	148·23	2	184·68
3	112·99	3	149·44	3	185·89
4	114·21	4	150·66	4	187·11
5	115·42	5	151·87	5	188·32
6	116·64	6	153·09	6	189·54
7	117·85	7	154·30	7	190·75
8	119·07	8	155·52	8	191·97
9	120·28	9	156·73	9	193·18
·100	121·50	·130	157·95	·160	194·40
1	122·71	1	159·16	1	195·61
2	123·93	2	160·38	2	196·83
3	125·14	3	161·59	3	198·04
4	126·36	4	162·81	4	199·26
5	127·57	5	164·02	5	200·47
6	128·79	6	165·24	6	201·69
7	130·00	7	166·45	7	202·90
8	131·22	8	167·67	8	204·12
9	132·43	9	168·88	9	205·33
·110	133·65	·140	170·10	·170	206·55
1	134·86	1	171·31	1	207·76
2	136·08	2	172·53	2	208·98
3	137·29	3	173·74	3	210·19
4	138·51	4	174·96	4	211·41
5	139·72	5	176·17	5	212·62
6	140·94	6	177·39	6	213·84
7	142·15	7	178·60	7	215·05
8	143·37	8	179·82	8	216·27
9	144·58	9	181·03	9	217·48

Ounces of Fine Metal per Ton of Ore of 2,000 Pounds.

Fine Metal.—Thousands of the unit of 10 grains.	One ton of ore will contain.—Ounces.	Fine Metal.—Thousands of the unit of 10 grains.	One ton of ore will contain.—Ounces.	Fine Metal.—Thousands of the unit of 10 grains.	One ton of ore will contain.—Ounces.
·180	218·70	·210	255·15	·240	291·60
1	219·91	1	256·36	1	292·81
2	221·13	2	257·58	2	294·03
3	222·34	3	258·79	3	295·24
4	223·56	4	260·01	4	296·46
5	224·77	5	261·22	5	297·67
6	225·99	6	262·44	6	398·89
7	227·20	7	263·65	7	300·10
8	228·42	8	264·87	8	301·32
9	229·63	9	266·08	9	302·53
·190	230·85	·220	267·30	·250	303·75
1	232·06	1	268·51	1	304·96
2	233·28	2	269·73	2	306·18
3	234·49	3	270·94	3	307·39
4	235·71	4	272·16	4	308·61
5	236·92	5	273·37	5	309·82
6	238·14	6	274·59	6	311·04
7	239·35	7	275·80	7	312·25
8	240·57	8	277·02	8	313·47
9	241·78	9	278·23	9	314·68
·200	243·00	·230	279·45	·260	315·90
1	244·21	1	280·66	1	317·11
2	245·43	2	281·88	2	318·33
3	246·64	3	283·09	3	319·54
4	247·86	4	284·31	4	320·76
5	249·07	5	285·52	5	321·97
6	250·29	6	286·74	6	323·19
7	251·50	7	287·95	7	324·40
8	252·72	8	289·17	8	325·62
9	253·93	9	290·38	9	326·83

Ounces of Fine Metal per Ton of Ore of 2,000 Pounds.

Fine Metal — Thousands of the unit of 10 grains.	One ton of ore will contain — Ounces.	Fine Metal — Thousands of the unit of 10 grains.	One ton of ore will contain — Ounces.	Fine Metal — Thousands of the unit of 10 grains.	One ton of ore will contain — Ounces.
·270	328·05	·300	364·50	·330	400·95
1	329·26	1	365·71	1	402·16
2	330·48	2	366·93	2	403·38
3	331·69	3	368·14	3	404·59
4	332·91	4	369·36	4	405·81
5	334·12	5	370·57	5	407·02
6	335·34	6	371·79	6	408·24
7	336·55	7	373·00	7	409·45
8	337·77	8	374·22	8	410·67
9	338·98	9	375·43	9	411·88
·280	340·20	·310	376·65	·340	413·10
1	341·41	1	377·86	1	414·31
2	342·63	2	379·08	2	415·53
3	343·84	3	380·29	3	416·74
4	345·06	4	381·51	4	417·96
5	346·27	5	382·72	5	419·17
6	347·49	6	383·94	6	420·39
7	348·70	7	385·15	7	421·60
8	349·92	8	386·37	8	422·82
9	351·13	9	387·58	9	424·03
·290	352·35	·320	388·80	·350	425·25
1	353·56	1	390·01	1	426·46
2	354·78	2	391·23	2	427·68
3	355·99	3	392·44	3	428·89
4	357·21	4	393·66	4	430·11
5	358·42	5	394·87	5	431·32
6	359·64	6	396·09	5	432·54
7	360·85	7	397·30	7	433·75
8	362·07	8	398·52	8	434·97
9	363·28	9	399·73	9	436·18

288   PROCESSES OF SILVER AND GOLD EXTRACTION.

Ounces of Fine Metal per Ton of Ore of 2,000 Pounds.

Fine Metal — Thousands of the unit of 10 grains.	One ton of ore will contain — Ounces.	Fine Metal — Thousands of the unit of 10 grains.	One ton of ore will contain — Ounces.	Fine Metal — Thousands of the unit of 10 grains.	One ton of ore will contain — Ounces.
·360	437·40	·390	473·85	·420	510·30
1	438·61	1	475·06	1	511·51
2	439·83	2	476·28	2	512·73
3	441·04	3	477·49	3	513·94
4	442·26	4	478·71	4	515·16
5	443·47	5	479·92	5	516·37
6	444·69	6	481·14	6	517·59
7	445·90	7	482·35	7	518·80
8	447·12	8	483·57	8	520·02
9	448·33	9	484·78	9	521·23
·370	449·55	·400	486·00	·430	522·45
1	450·76	1	487·21	1	523·66
2	451·98	2	488·43	2	524·88
3	453·19	3	489·64	3	526·09
4	454·41	4	490·86	4	527·31
5	455·62	5	492·07	5	528·52
6	456·84	6	493·29	6	529·74
7	458·05	7	494·50	7	530·95
8	459·27	8	495·72	8	532·17
9	460·48	9	496·93	9	533·38
·380	461·70	·410	498·15	·440	534·60
1	462·91	1	499·36	1	535·81
2	464·13	2	500·58	2	537·03
3	465·34	3	501·79	3	538·24
4	466·56	4	503·01	4	539·46
5	467·77	5	504·22	5	540·67
6	468·99	6	505·44	6	541·89
7	470·20	7	506·65	7	543·10
8	471·42	8	507·87	8	544·32
9	472·63	9	509·08	9	545·53

PROCESSES OF SILVER AND GOLD EXTRACTION.

Ounces of Fine Metal per Ton of Ore of 2,000 Pounds.

Fine Metal — Thousands of the unit of 10 grains.	One ton of ore will contain — Ounces.	Fine Metal — Thousands of the unit of 10 grains.	One ton of ore will contain — Ounces.	Fine Metal — Thousands of the unit of 10 grains.	One ton of ore will contain — Ounces.
·450	546·75	·480	583·20	·510	619·65
1	547·96	1	584·41	1	620·86
2	549·18	2	585·63	2	622·08
3	550·39	3	586·84	3	623·29
4	551·61	4	588·06	4	624·51
5	552·82	5	589·27	5	625·72
6	554·04	6	590·49	6	626·94
7	555·25	7	591·70	7	628·15
8	556·47	8	592·92	8	629·37
9	557·68	9	594·13	9	630·58
·460	558·90	·490	595·35	·520	631·80
1	560·11	1	596·56	1	633·01
2	561·33	2	597·78	2	634·23
3	562·54	3	598·99	3	635·44
4	563·76	4	600·21	4	636·66
5	564·97	5	601·42	5	637·87
6	566·19	6	602·64	6	639·09
7	567·40	7	603·85	7	640·30
8	568·62	8	605·07	8	641·52
9	569·83	9	606·28	9	642·73
·470	571·05	·500	607·50	·530	643·95
1	572·26	1	608·71	1	645·16
2	573·48	2	609·93	2	646·38
3	574·69	3	611·14	3	647·59
4	575·91	4	612·36	4	648·81
5	577·12	5	613·57	5	650·02
6	578·34	6	614·79	6	651·24
7	579·55	7	616·00	7	652·45
8	580·77	8	617·22	8	653·67
9	581·98	9	618·43	9	654·88

Ounces of Fine Metal per Ton of Ore of 2,000 Pounds.

Fine Metal — Thousands of the unit of 10 grains.	One ton of ore will contain — Ounces.	Fine Metal — Thousands of the unit of 10 grains.	One ton of ore will contain — Ounces.	Fine Metal — Thousands of the unit of 10 grains.	One ton of ore will contain — Ounces.
·540	656·10	·570	692·55	·600	729·00
1	657·31	1	693·76	1	730·21
2	658·53	2	694·98	2	731·43
3	659·74	3	696·19	3	732·64
4	660·96	4	697·41	4	733·86
5	662·17	5	698·62	5	735·07
6	663·39	6	699·84	6	736·29
7	664·60	7	701·05	7	737·50
8	665·82	8	702·27	8	738·72
9	667·03	9	703·48	9	739·93
·550	668·25	·580	704·70	·610	741·15
1	669·46	1	705·91	1	742·36
2	670·68	2	707·13	2	743·58
3	671·89	3	708·34	3	744·79
4	673·11	4	709·56	4	746·01
5	674·32	5	710·77	5	747·22
6	675·54	6	711·99	6	748·44
7	676·75	7	713·20	7	749·65
8	677·97	8	714·42	8	650·87
9	679·18	9	715·63	9	752·08
·560	680·40	·590	716·85	·620	753·30
1	681·61	1	718·06	1	754·52
2	682·83	2	719·28	2	755·74
3	684·04	3	720·49	3	756·95
4	685·26	4	721·71	4	758·17
5	686·47	5	722·92	5	759·38
6	687·69	6	724·14	6	760·59
7	688·90	7	725·35	7	761·80
8	690·12	8	726·57	8	763·02
9	691·33	9	727·78	9	764·23

PROCESSES OF SILVER AND GOLD EXTRACTION. 291

Ounces of Fine Metal per Ton of Ore of 2,000 Pounds.

Fine Metal — Thousands of the unit of 10 grains.	One ton of ore will contain — Ounces.	Fine Metal — Thousands of the unit of 10 grains.	One ton of ore will contain — Ounces.	Fine Metal — Thousands of the unit of 10 grains.	One ton of ore will contain — Ounces.
·630	765·45	·660	801·90	·690	838·35
1	766·66	1	803·11	1	839·56
2	767·88	2	804·33	2	840·78
3	769·09	3	805·54	3	841·99
4	770·31	4	806·76	4	843·21
5	771·52	5	807·97	5	844·42
6	772·74	6	809·19	6	845·64
7	773·95	7	810·40	7	846·85
8	775·17	8	811·62	8	848·07
9	776·38	9	812·83	9	849·28
·640	777·60	·670	814·05	·700	850·50
1	778·82	1	815·26	1	851·71
2	780·04	2	816·48	2	852·93
3	781·25	3	817·69	3	854·14
4	782·47	4	818·91	4	855·36
5	783·68	5	820·12	5	856·57
6	784·90	6	821·34	6	857·79
7	786·11	7	822·55	7	859·00
8	787·33	8	823·77	8	860·22
9	788·54	9	824·98	9	861·43
·650	789·75	·680	826·20	·710	862·65
1	790·96	1	827·41	1	863·86
2	792·18	2	828·63	2	865·08
3	793·39	3	829·84	3	866·29
4	794·61	4	831·06	4	867·51
5	795·82	5	832·27	5	868·72
6	797·04	6	833·49	6	869·94
7	798·25	7	834·70	7	871·15
8	799·47	8	835·92	8	872·37
9	800·68	9	837·13	9	873·58

PROCESSES OF SILVER AND GOLD EXTRACTION.

Ounces of **Fine Metal** *per* **Ton** *of* **Ore** *of* 2,000 *Pounds.*

Fine Metal — Thousands of the unit of 10 grains.	One ton of ore will contain — Ounces.	Fine Metal — Thousands of the unit of 10 grains.	One ton of ore will contain — Ounces.	Fine Metal — Thousands of the unit of 10 grains.	One ton of ore will contain — Ounces.
·720	874·80	·750	911·25	·780	947·70
1	876·01	1	912·46	1	948·91
2	877·23	2	913·68	2	950·13
3	878·44	3	914·89	3	951·34
4	879·66	4	916·11	4	952·56
5	880·87	5	917·32	5	953·77
6	882·09	6	918·54	6	954·99
7	883·30	7	919·75	7	956·20
8	884·52	8	920·97	8	957·42
9	885·73	9	922·18	9	958·63
·730	886·95	·760	923·40	·790	959·85
1	888·16	1	924·61	1	961·06
2	889·38	2	925·83	2	962·28
3	890·59	3	927·04	3	963·49
4	891·81	4	928·26	4	964·71
5	893·02	5	929·47	5	965·92
6	894·24	6	930·69	6	967·14
7	895·45	7	931·90	7	968·35
8	896·67	8	933·12	8	969·57
9	897·88	9	934·33	9	970·78
·740	899·10	·770	935·55	·800	972·00
1	900·31	1	936·76	1	973·21
2	901·53	2	937·98	2	974·43
3	902·74	3	939·19	3	975·64
4	903·96	4	940·41	4	976·86
5	905·17	5	941·62	5	978·07
6	906·39	6	942·83	6	979·29
7	907·60	7	944·05	7	980·50
8	908·82	8	945·27	8	981·72
9	910·03	9	946·48	9	982·93

PROCESSES OF SILVER AND GOLD EXTRACTION.

Ounces of Fine Metal per Ton of Ore of 2,000 Pounds.

Fine Metal — Thousands of the unit of 10 grains.	One ton of ore will contain — Ounces.	Fine Metal — Thousands of the unit of 10 grains.	One ton of ore will contain — Ounces.	Fine Metal — Thousands of the unit of 10 grains.	One ton of ore will contain — Ounces.
·810	984·15	·840	1020·60	·870	1057·05
1	985·36	1	1021·81	1	1058·26
2	986·58	2	1023·03	2	1059·48
3	987·79	3	1024·24	3	1060·69
4	989·01	4	1025·46	4	1061·91
5	990·22	5	1026·67	5	1063·12
6	991·44	6	1027·89	6	1064·34
7	992·65	7	1029·10	7	1065·55
8	993·87	8	1030·32	8	1066·77
9	995·08	9	1031·53	9	1067·98
·820	996·30	·850	1032·75	·880	1069·20
1	997·51	1	1033·96	1	1070·41
2	998·73	2	1035·18	2	1071·63
3	999·94	3	1036·39	3	1072·84
4	1001·16	4	1037·61	4	1074·06
5	1002·37	5	1038·82	5	1075·27
6	1003·59	6	1040·04	6	1076·49
7	1004·80	7	1041·25	7	1077·70
8	1006·02	8	1042·47	8	1078·92
9	1007·23	9	1043·68	9	1080·13
·830	1008·45	·860	1044·90	·890	1081·35
1	1009·66	1	1046·11	1	1082·56
2	1010·88	2	1047·33	2	1083·78
3	1012·09	3	1048·54	3	1084·99
4	1013·31	4	1049·76	4	1086·21
5	1014·52	5	1050·97	5	1087·42
6	1015·74	6	1052·19	6	1088·64
7	1016·95	7	1053·40	7	1089·85
8	1018·17	8	1054·62	8	1091·07
9	1019·38	9	1055·83	9	1092·28

Ounces of Fine Metal per Ton of Ore of 2,000 Pounds.

Fine Metal — Thousands of the unit of 10 grains.	One ton of ore will contain — Ounces.	Fine Metal — Thousands of the unit of 10 grains.	One ton of ore will contain — Ounces.	Fine Metal — Thousands of the unit of 10 grains.	One ton of ore will contain — Ounces.
·900	1093·50	·930	1129·95	·960	1166·40
1	1094·71	1	1131·16	1	1167·61
2	1095·93	2	1132·38	2	1168·83
3	1097·14	3	1133·59	3	1170·04
4	1098·36	4	1134·81	4	1171·26
5	1099·57	5	1136·02	5	1172·47
6	1100·79	6	1137·24	6	1173·69
7	1102·00	7	1138·45	7	1174·90
8	1103·22	8	1139·67	8	1176·12
9	1104·43	9	1140·88	9	1177·33
·910	1105·65	·940	1142·10	·970	1178·55
1	1106·86	1	1143·31	1	1179·76
2	1108·08	2	1144·53	2	1180·98
3	1109·29	3	1145·75	3	1182·19
4	1110·51	4	1146·96	4	1183·41
5	1111·72	5	1148·17	5	1184·62
6	1112·94	6	1149·39	6	1185·84
7	1114·15	7	1150·60	7	1187·05
8	1115·37	8	1151·82	8	1188·27
9	1116·58	9	1153·03	9	1189·48
·920	1117·80	·950	1154·25	·980	1190·70
1	1119·01	1	1155·46	1	1191·91
2	1120·23	2	1156·68	2	1193·13
3	1121·44	3	1157·89	3	1194·34
4	1122·66	4	1159·11	4	1195·56
5	1123·87	5	1160·32	5	1196·77
6	1125·09	6	1161·54	6	1197·99
7	1126·30	7	1162·75	7	1199·20
8	1127·52	8	1163·97	8	1200·42
9	1128·73	9	1165·18	9	1201·63

Ounces of Fine Metal per Ton of Ore of 2,000 *Pounds.*

Fine Metal — Thousands of the unit of 10 grains.	One ton of ore will contain — Ounces.	Fine Metal — Thousands of the unit of 10 grains.	One ton of ore will contain — Ounces.	Fine Metal — Thousands of the unit of 10 grains.	One ton of ore will contain — Ounces.
·990	1202·85	·995	1208·92	1·000	1215·00
1	1204·06	6	1210·14	2·000	2430·00
2	1205·28	7	1211·35	3·000	3645·00
3	1206·49	8	1212·57	4·000	4860·00
4	1207·71	9	1213·78	5·000	6075·00

TABLE II

SHOWING THE VALUE OF SILVER PER OUNCE TROY AS ALLOYED IN THE BAR WHEN THE STANDARD IS EXPRESSED IN THOUSANDTHS.

Value of Silver per Ounce Troy at different Fineness.

Thousandths Fine.	Dollars.	Cents.	Thousandths Fine.	Dollars.	Cents.	Thousandths Fine.	Dollars.	Cents.
½		00·06	·030		03·88	·060		07·76
·001		00·13	1		04·01	1		07·89
2		00·26	2		04·14	2		08·02
3		00·39	3		04·27	3		08·15
4		09·52	4		04·40	4		08·28
5		00·65	5		04·53	5		08·41
6		00·78	6		04·66	6		08·53
7		00·91	7		04·78	7		08·66
8		01·04	8		04·91	8		08·79
9		01·17	9		05·04	9		08·92
·010		01·29	·040		05·17	·070		09·05
1		01·42	1		05·30	1		09·18
2		01·55	2		05·43	2		09·31
3		01·68	3		05·56	3		09·44
4		01·81	4		05·69	4		09·57
5		01·94	5		05·82	5		09·70
6		02·07	6		05·95	6		09·83
7		02·20	7		06·08	7		09·96
8		02·33	8		06·21	8		10·09
9		02·46	9		06·34	9		10·21
·020		02·58	·050		06·46	·080		10·34
1		02·71	1		06·59	1		10·47
2		02·83	2		06·72	2		10·60
3		02·96	3		06·85	3		10·73
4		03·09	4		06·98	4		10·86
5		03·22	5		07·11	5		10·99
6		03·35	6		07·24	6		11·12
7		03·49	7		07·37	7		11·25
8		03·62	8		07·50	8		11·38
9		03·75	9		07·63	9		11·51

Value of Silver per Ounce Troy at different Fineness.

Thousandths Fine	Dollars	Cents	Thousandths Fine	Dollars	Cents	Thousandths Fine	Dollars	Cents
·090		11·64	·120		15·52	·150		19·39
1		11·77	1		15·64	1		19·52
2		11·89	2		15·77	2		19·65
3		12·02	3		15·90	3		19·78
4		12·15	4		16·03	4		19·91
5		12·28	5		16·16	5		20·04
6		12·41	6		16·29	6		20·17
7		12·54	7		16·42	7		20·30
8		12·67	8		16·55	8		20·43
9		12·80	9		16·68	9		20·56
·100		12·93	·130		16·81	·160		20·69
1		13·06	1		16·94	1		20·82
2		13·19	2		17·07	2		20·95
3		13·32	3		17·20	3		21·08
4		13·45	4		17·33	4		21·21
5		13·58	5		17·45	5		21·33
6		13·71	6		17·58	6		21·46
7		13·83	7		17·71	7		21·59
8		13·96	8		17·84	8		21·72
9		14·09	9		17·97	9		21·85
·110		14·22	·140		18·10	·170		21·98
1		14·35	1		18·23	1		22·11
2		14·48	2		18·36	2		22·24
3		14·61	3		18·49	3		22·37
4		14·74	4		18·62	4		22·50
5		14·87	5		18·75	5		22·63
6		15·00	6		18·88	6		22·75
7		15·13	7		19·01	7		22·88
8		15·26	8		19·14	8		23·01
9		15·39	9		19·26	9		23·14

Value of Silver per Ounce Troy at different Fineness.

Thousandths Fine	Dollars	Cents	Thousandths Fine	Dollars	Cents	Thousandths Fine	Dollars	Cents
·180		23·27	·210		27·15	·240		31·03
1		23·40	1		27·28	1		31·16
2		23·53	2		27·41	2		31·29
3		23·66	3		27·54	3		31·42
4		23·79	4		27·67	4		31·55
5		23·92	5		27·80	5		31·68
6		24·05	6		27·93	6		31·81
7		24·18	7		28·06	7		31·94
8		24·31	8		28·19	8		32·06
9		24·44	9		28·32	9		32·19
·190		24·57	·220		28·44	·250		32·32
1		24·69	1		28·57	1		32·45
2		24·82	2		28·70	2		32·58
3		24·95	3		28·83	3		32·71
4		25·08	4		28·96	4		32·84
5		25·21	5		29·09	5		32·97
6		25·34	6		29·22	6		33·10
7		25·47	7		29·35	7		33·23
8		25·60	8		29·48	8		33·36
9		25·73	9		29·61	9		33·49
·200		25·86	·230		29·74	·260		33·62
1		25·99	1		29·87	1		33·75
2		26·12	2		30·00	2		33·87
3		26·25	3		30·13	3		34·00
4		26·38	4		30·26	4		34·13
5		26·50	5		30·38	5		34·26
6		26·63	6		30·51	6		34·39
7		26·76	7		30·64	7		34·52
8		26·89	8		30·77	8		34·65
9		27·02	9		30·90	9		34·78

PROCESSES OF SILVER AND GOLD EXTRACTION. 301

Value of Silver per Ounce Troy at different Fineness.

Thousandths Fine	Dollars	Cents	Thousandths Fine	Dollars	Cents	Thousandths Fine	Dollars	Cents
·270		34·91	·300		38·79	·330		42·67
1		35·04	1		38·92	1		42·80
2		35·17	2		39·05	2		42·93
3		35·30	3		39·18	3		43·05
4		35·43	4		39·30	4		43·18
5		35·56	5		39·43	5		43·31
6		35·69	6		39·56	6		43·44
7		35·81	7		39·69	7		43·57
8		35·94	8		39·82	8		43·70
9		36·07	9		39·95	9		43·83
·280		36·20	·310		40·08	·340		43·96
1		36·33	1		40·21	1		44·09
2		36·46	2		40·34	2		44·22
3		36·59	3		40·47	3		44·35
4		36·72	4		40·60	4		44·48
5		36·85	5		40·73	5		44·61
6		36·98	6		40·86	6		44·74
7		37·11	7		40·98	7		44·86
8		37·24	8		41·11	8		44·99
9		37·37	9		41·24	9		45·12
·290		37·50	·320		41·37	·350		45·25
1		37·63	1		41·50	1		45·38
2		37·76	2		41·63	2		45·51
3		37·88	3		41·76	3		45·64
4		38·01	4		41·89	4		45·77
5		38·14	5		42·02	5		45·90
6		38·27	6		42·15	6		46·03
7		38·40	7		42·28	7		46·16
8		38·53	8		42·41	8		46·29
9		38·66	9		42·54	9		46·42

Value of Silver per Ounce Troy at different Fineness.

Thousandths Fine.	Dollars.	Cents.	Thousandths Fine.	Dollars.	Cents.	Thousandths Fine.	Dollars.	Cents.
·360		46·55	·390		50·42	·420		54·30
1		46·67	1		50·55	1		54·43
2		46·80	2		50·68	2		54·56
3		46·93	3		50·81	3		54·69
4		47·06	4		50·94	4		54·82
5		47·19	5		51·07	5		54·95
6		47·32	6		51·20	6		55·08
7		47·45	7		51·33	7		55·21
8		47·58	8		51·46	8		55·34
9		47·71	9		51·59	9		55·47
·370		47·84	·400		51·72	·430		55·60
1		47·97	1		51·85	1		55·73
2		48·10	2		51·98	2		55·85
3		48·23	3		52·11	3		55·98
4		48·36	4		52·23	4		56·11
5		48·48	5		52·36	5		56·24
6		48·61	6		52·49	6		56·37
7		48·74	7		52·62	7		56·50
8		48·87	8		52·75	8		56·63
9		49·00	9		52·88	9		56·76
·380		49·13	·410		53·01	·440		56·89
1		49·26	1		53·14	1		57·02
2		49·39	2		53·27	2		57·15
3		49·52	3		53·40	3		57·28
4		49·65	4		53·53	4		57·41
5		49·78	5		53·66	5		57·54
6		49·91	6		53·79	6		57·66
7		50·04	7		53·92	7		57·79
8		50·17	8		54·04	8		57·92
9		50·29	9		54·17	9		58·05

PROCESSES OF SILVER AND GOLD EXTRACTION.   303

Value of Silver per Ounce Troy at different Fineness.

Thousandths Fine	Dollars	Cents	Thousandths Fine	Dollars	Cents	Thousandths Fine	Dollars	Cents
·450		58·18	·480		62·06	·510		65·94
1		58·31	1		62·19	1		66·07
2		58·44	2		62·32	2		66·20
3		58·57	3		62·45	3		66·33
4		58·70	4		62·58	4		66·46
5		58·83	5		62·71	5		66·59
6		58·96	6		62·84	6		66·72
7		59·09	7		62·97	7		66·84
8		59·22	8		63·09	8		66·97
9		59·35	9		63·22	9		67·10
·460		59·47	·490		63·35	·520		67·23
1		59·60	1		63·48	1		67·36
2		59·73	2		63·61	2		67·49
3		59·86	3		63·74	3		67·62
4		59·99	4		63·87	4		67·75
5		60·12	5		64·00	5		67·88
6		60·25	6		64·13	6		68·01
7		60·38	7		64·26	7		68·14
8		60·51	8		64·39	8		68·27
9		60·64	9		64·52	9		68·40
·470		60·77	·500		64·65	·530		68·53
1		60·90	1		64·78	1		68·65
2		61·02	2		64·91	2		68·78
3		61·15	3		65·03	3		68·91
4		61·28	4		65·16	4		69·04
5		61·41	5		65·29	5		69·17
6		61·54	6		65·42	6		69·30
7		61·67	7		65·55	7		69·43
8		61·80	8		65·68	8		69·56
9		61·93	9		65·81	9		69·69

Value of Silver per Ounce Troy at different Fineness.

Thousandths Fine.	Dollars.	Cents.	Thousandths Fine.	Dollars.	Cents.	Thousandths Fine.	Dollars.	Cents.
·540		69·82	·570		73·69	·600		77·58
1		69·95	1		73·82	1		77·71
2		70·08	2		73·95	2		77·83
3		70·21	3		74·08	3		77·96
4		70·34	4		74·21	4		78·09
5		70·46	5		74·34	5		78·22
6		70·59	6		74·47	6		78·35
7		70·72	7		74·60	7		78·48
8		70·85	8		74·73	8		78·61
9		70·98	9		74·86	9		78·74
·550		71·11	·580		74·99	·610		78·87
1		71·24	1		75·12	1		79·00
2		71·37	2		75·25	2		79·13
3		71·50	3		75·38	3		79·26
4		71·63	4		75·51	4		79·39
5		71·76	5		75·64	5		79·52
6		71·89	6		75·77	6		79·64
7		72·02	7		75·90	7		79·77
8		72·15	8		76·02	8		79·90
9		72·27	9		76·15	9		80·03
·560		72·40	·590		76·28	·620		80·16
1		72·53	1		76·41	1		80·29
2		72·66	2		76·54	2		80·42
3		72·79	3		76·67	3		80·55
4		72·92	4		76·80	4		80·68
5		73·05	5		76·93	5		80·81
6		73·18	6		77·06	6		80·94
7		73·31	7		77·19	7		81·07
8		73·44	8		77·32	8		81·20
9		73·56	9		77·45	9		81·32

PROCESSES OF SILVER AND GOLD EXTRACTION. 305

Value of Silver per Ounce Troy at different Fineness.

Thousandths Fine	Dollars	Cents	Thousandths Fine	Dollars	Cents	Thousandths Fine	Dollars	Cents
·630		81·45	·660		85·33	·690		89·21
1		81·58	1		85·46	1		89·34
2		81·71	2		85·59	2		89·47
3		81·84	3		85·72	3		89·60
4		81·97	4		85·85	4		89·73
5		82·10	5		85·98	5		89·86
6		82·23	6		86·11	6		89·99
7		82·36	7		86·24	7		90·12
8		82·49	8		86·37	8		90·25
9		82·62	9		86·50	9		90·38
·640		82·75	·670		86·63	·700		90·51
1		82·88	1		86·76	1		90·63
2		83·01	2		86·88	2		90·76
3		83·14	3		87·01	3		90·89
4		83·27	4		87·14	4		91·02
5		83·39	5		87·27	5		91·15
6		83·52	6		87·40	6		91·28
7		83·65	7		87·53	7		91·41
8		83·78	8		87·66	8		91·54
9		83·91	9		87·79	9		91·67
·650		84·04	·680		87·92	·710		91·80
1		84·17	1		88·05	1		91·93
2		84·30	2		88·18	2		92·06
3		84·43	3		88·31	3		92·19
4		84·56	4		88·44	4		92·32
5		84·69	5		88·57	5		92·45
6		84·82	6		88·69	6		92·57
7		84·94	7		88·82	7		92·70
8		85·07	8		88·95	8		92·83
9		85·20	9		89·08	9		92·96

Value of Silver per Ounce Troy at different Fineness.

Thousandths Fine	Dollars	Cents	Thousandths Fine	Dollars	Cents	Thousandths Fine	Dollars	Cents
·720		93·09	·750		96·97	·780	1	00·85
1		93·22	1		97·10	1	1	00·98
2		93·35	2		97·23	2	1	01·11
3		93·48	3		97·36	3	1	01·24
4		93·61	4		97·49	4	1	01·37
5		93·74	5		97·62	5	1	01·49
6		93·87	6		97·75	6	1	01·62
7		94·00	7		97·87	7	1	01·75
8		94·13	8		98·00	8	1	01·88
9		94·25	9		98·13	9	1	02·01
·730		94·38	·760		98·26	·790	1	02·14
1		94·51	1		98·39	1	1	02·27
2		94·64	2		98·52	2	1	02·40
3		94·77	3		98·65	3	1	02·53
4		94·90	4		98·78	4	1	02·66
5		95·03	5		98·91	5	1	02·79
6		95·16	6		99·04	6	1	02·92
7		95·29	7		99·17	7	1	03·05
8		95·42	8		99·30	8	1	03·18
9		95·55	9		99·43	9	1	03·31
·740		95·68	·770		99·56	·800	1	03·43
1		95·81	·1		99·68	1	1	03·56
2		95·94	2		99·81	2	1	03·69
3		96·06	3		99·94	3	1	03·82
4		96·19	4	1	00·07	4	1	03·95
5		96·32	5	1	00·20	5	1	04·07
6		96·45	6	1	00·33	6	1	04·21
7		96·58	7	1	00·46	7	1	04·34
8		96·71	8	1	00·59	8	1	04·47
9		96·84	9	1	00·72	9	1	04·60

Value of Silver per Ounce Troy at different Fineness.

Thousandths Fine	Dollars	Cents	Thousandths Fine	Dollars	Cents	Thousandths Fine	Dollars	Cents
·810	1	04·73	·840	1	08·61	·870	1	12·48
1	1	04·86	1	1	08·74	1	1	12·61
2	1	04·99	2	1	08·86	2	1	12·74
3	1	05·12	3	1	08·99	3	1	12·87
4	1	05·24	4	1	09·12	4	1	13·00
5	1	05·37	5	1	09·25	5	1	13·13
6	1	05·50	6	1	09·38	6	1	13·26
7	1	05·63	7	1	09·51	7	1	13·39
8	1	05·76	8	1	09·64	8	1	13·52
9	1	05·89	9	1	09·77	9	1	13·65
·820	1	06·02	·850	1	09·90	·880	1	13·78
1	1	06·15	1	1	10·03	1	1	13·91
2	1	06·28	2	1	10·16	2	1	14·04
3	1	06·41	3	1	10·29	3	1	14·17
4	1	06·54	4	1	10·42	4	1	14·29
5	1	06·67	5	1	10·55	5	1	14·42
6	1	06·80	6	1	10·67	6	1	14·55
7	1	06·93	7	1	10·80	7	1	14·68
8	1	07·05	8	1	10·93	8	1	14·81
9	1	07·18	9	1	11·06	9	1	14·94
·830	1	07·31	·860	1	11·19	·890	1	15·07
1	1	07·44	1	1	11·32	1	1	15·20
2	1	07·57	2	1	11·45	2	1	15·33
3	1	07·70	3	1	11·58	3	1	15·46
4	1	07·83	4	1	11·71	4	1	15·59
5	1	07·96	5	1	11·84	5	1	15·72
6	1	08·09	6	1	11·97	6	1	15·85
7	1	08·22	7	1	12·10	7	1	15·98
8	1	08·35	8	1	12·23	8	1	16·11
9	1	08·48	9	1	12·36	9	1	16·23

PROCESSES OF SILVER AND GOLD EXTRACTION.

Value of Silver per Ounce Troy at different Fineness.

Thousandths Fine.	Dollars.	Cents.	Thousandths Fine.	Dollars.	Cents.	Thousandths Fine.	Dollars.	Cents.
·900	1	16·36	·930	1	20·24	·960	1	24·12
1	1	16·49	1	1	20·37	1	1	24·25
2	1	16·62	2	1	20·50	2	1	24·38
3	1	16·75	3	1	20·63	3	1	24·51
4	1	16·88	4	1	20·76	4	1	24·64
5	1	17·01	5	1	20·89	5	1	24·77
6	1	17·14	6	1	21·02	6	1	24·90
7	1	17·27	7	1	21·15	7	1	25·03
8	1	17·40	8	1	21·28	8	1	25·16
9	1	17·53	9	1	21·41	9	1	25·28
·910	1	17·66	·940	1	21·54	·970	1	25·41
1	1	17·79	1	1	21·66	1	1	25·54
2	1	17·92	2	1	21·79	2	1	25·67
3	1	18·05	3	1	21·92	3	1	25·80
4	1	18·17	4	1	22·05	4	1	25·93
5	1	18·30	5	1	22·18	5	1	26·06
6	1	18·43	6	1	22·31	6	1	26·19
7	1	18·56	7	1	22·44	7	1	26·32
8	1	18·69	8	1	22·57	8	1	26·45
9	1	18·82	9	1	22·70	9	1	26·57
·920	1	18·95	·950	1	22·83	·980	1	26·71
1	1	19·08	1	1	22·96	1	1	26·84
2	1	19·21	2	1	23·09	2	1	26·97
3	1	19·34	3	1	23·22	3	1	27·09
4	1	19·47	4	1	23·35	4	1	27·22
5	1	19·60	5	1	23·47	5	1	27·35
6	1	19·72	6	1	23·60	6	1	27·48
7	1	19·85	7	1	23·73	7	1	27·61
8	1	19·98	8	1	23·86	8	1	27·74
9	1	20·11	9	1	23·99	9	1	27·87

Value of Silver per Ounce Troy at different Fineness.

Thousandths Fine	Dollars	Cents	Thousandths Fine	Dollars	Cents	Thousandths Fine	Dollars	Cents
·990	1	28·00	·994	1	28·52	·998	1	29·03
1	1	28·13	5	1	28·65	9	1	29·16
2	1	28·26	6	1	28·78	1000	1	29·29
3	1	28·39	7	1	28·90			

TABLE III

SHOWING THE VALUE OF GOLD PER OUNCE TROY AS IT IS ALLOYED IN THE BAR.

Value of Gold per Ounce Troy at different Fineness.

Thousandths Fine	Dollars	Cents	Thousandths Fine	Dollars	Cents	Thousandths Fine	Dollars	Cents
½		01·03	·030		62·02	·060	1	24·03
·001		02·07	1		64·08	1	1	26·10
2		04·13	2		66·15	2	1	28·17
3		06·20	3		68·22	3	1	30·23
4		08·27	4		70·28	4	1	32·30
5		10·34	5		72·35	5	1	34·37
6		12·40	6		74·42	6	1	36·43
7		14·47	7		76·49	7	1	38·50
8		16·54	8		78·55	8	1	40·57
9		18·60	9		80·62	9	1	42·64
·010		20·67	·040		82·69	·070	1	44·70
1		22·74	1		84·75	1	1	46·77
2		24·81	2		86·82	2	1	48·84
3		26·87	3		88·89	3	1	50·90
4		28·94	4		90·96	4	1	52·97
5		31·01	5		93·02	5	1	55·04
6		33·07	6		95·09	6	1	57·11
7		35·14	7		97·16	7	1	59·17
8		37·21	8		99·22	8	1	61·24
9		39·28	9	1	01·29	9	1	63·31
·020		41·34	·050	1	03·36	·080	1	65·37
1		43·41	1	1	05·43	1	1	67·44
2		45·48	2	1	07·49	2	1	69·51
3		47·55	3	1	09·56	3	1	71·58
4		49·61	4	1	11·63	4	1	73·64
5		51·68	5	1	13·70	5	1	75·71
6		53·75	6	1	15·76	6	1	77·78
7		55·81	7	1	17·83	7	1	79·84
8		57·88	8	1	19·90	8	1	81·91
9		59·95	9	1	21·96	9	1	83·98

PROCESSES OF SILVER AND GOLD EXTRACTION. 313

Value of Gold per Ounce Troy at different Fineness.

Thousandths Fine.	Dollars.	Cents.	Thousandths Fine.	Dollars.	Cents.	Thousandths Fine.	Dollars.	Cents.
·090	1	86·05	·120	2	48·06	·150	3	10·08
1	1	88·11	1	2	50·13	1	3	12·14
2	1	90·18	2	2	52·20	2	3	14·21
3	1	92·25	3	2	54·26	3	3	16·28
4	1	94·32	4	2	56·33	4	3	18·35
5	1	96·38	5	2	58·40	5	3	20·41
6	1	98·45	6	2	60·46	6	3	22·48
7	2	00·52	7	2	62·53	7	3	24·55
8	2	02·58	8	2	64·60	8	3	26·61
9	2	04·65	9	2	66·67	9	3	28·68
·100	2	06·72	·130	2	68·73	·160	3	30·75
1	2	08·79	1	2	70·80	1	3	32·82
2	2	10·85	2	2	72·87	2	3	34·88
3	2	12·92	3	2	74·94	3	3	36·95
4	2	14·99	4	2	77·00	4	3	39·02
5	2	17·05	5	2	79·07	5	3	41·09
6	2	19·12	6	2	81·14	6	3	43·15
7	2	21·19	7	2	83·20	7	3	45·22
8	2	23·26	8	2	85·27	8	3	47·29
9	2	25·32	9	2	87·34	9	3	49·35
·110	2	27·39	·140	2	89·41	·170	3	51·42
1	2	29·46	1	2	91·47	1	3	53·49
2	2	31·52	2	2	93·54	2	3	55·56
3	2	33·59	3	2	95·61	3	3	57·62
4	2	35·66	4	2	97·67	4	3	59·69
5	2	37·73	5	2	99·74	5	3	61·76
6	2	39·79	6	3	01·81	6	3	63·82
7	2	41·86	7	3	03·88	7	3	65·89
8	2	43·93	8	3	05·94	8	3	67·96
9	2	45·99	9	3	08·01	9	3	70·03

Value of Gold per Ounce Troy at *different Fineness.*

Thousandths Fine.	Dollars.	Cents.	Thousandths Fine.	Dollars.	Cents.	Thousandths Fine.	Dollars.	Cents.
·180	3	72·09	·210	4	34·11	·240	4	96·12
1	3	74·16	1	4	36·18	1	4	98·19
2	3	76·23	2	4	38·24	2	5	00·26
3	3	78·29	3	4	40·31	3	5	02·33
4	3	80·36	4	4	42·38	4	5	04·39
5	3	82·43	5	4	44·44	5	5	06·46
6	3	84·50	6	4	46·51	6	5	08·53
7	3	86·56	7	4	48·58	7	5	10·59
8	3	88·63	8	4	50·65	8	5	12·66
9	3	90·70	9	4	52·71	9	5	14·73
·190	3	92·76	·220	4	54·78	·250	5	16·80
1	3	94·83	1	4	56·85	1	5	18·86
2	3	96·90	2	4	58·91	2	5	20·93
3	3	98·97	3	4	60·98	3	5	23·00
4	4	01·03	4	4	63·05	4	5	25·06
5	4	03·10	5	4	65·12	5	5	27·13
6	4	05·17	6	4	67·18	6	5	29·20
7	4	07·24	7	4	69·25	7	5	31·27
8	4	09·30	8	4	71·32	8	5	33·33
9	4	11·37	9	4	73·39	9	5	35·40
·200	4	13·44	·230	4	75·45	·260	5	37·47
1	4	15·50	1	4	77·52	1	5	39·53
2	4	17·57	2	4	79·59	2	5	41·60
3	4	19·64	3	4	81·65	3	5	43·67
4	4	21·71	4	4	83·72	4	5	45·74
5	4	23·78	5	4	85·79	5	5	47·80
6	4	25·84	6	4	87·86	6	5	49·87
7	4	27·91	7	4	89·92	7	5	51·94
8	4	29·97	8	4	91·99	8	5	54·01
9	4	32·04	9	4	94·06	9	5	56·07

Value of Gold per Ounce Troy at different Fineness.

Thousandths Fine	Dollars	Cents	Thousandths Fine	Dollars	Cents	Thousandths Fine	Dollars	Cents
·270	5	58·14	·300	6	20·16	·330	6	82·17
1	5	60·21	1	6	22·22	1	6	84·24
2	5	62·27	2	6	24·29	2	6	86·30
3	5	64·34	3	6	26·36	3	6	88·37
4	5	66·41	4	6	28·42	4	6	90·44
5	5	68·48	5	6	30·49	5	6	92·51
6	5	70·54	6	6	32·56	6	6	94·57
7	5	72·61	7	6	34·63	7	6	96·64
8	5	74·68	8	6	36·69	8	6	98·71
9	5	76·74	9	6	38·76	9	7	00·78
·280	5	78·81	·310	6	40·83	·340	7	02·84
1	5	80·88	1	6	42·89	1	7	04·91
2	5	82·95	2	6	44·96	2	7	06·98
3	5	85·01	3	6	47·03	3	7	09·04
4	5	87·08	4	6	49·10	4	7	11·11
5	5	89·15	5	6	51·16	5	7	13·18
6	5	91·21	6	6	53·23	6	7	15·25
7	5	93·28	7	6	55·30	7	7	17·31
8	5	95·35	8	6	57·36	8	7	19·38
9	5	97·42	9	6	59·43	9	7	21·45
·290	5	99·48	·320	6	61·50	·350	7	23·51
1	6	01·55	1	6	63·57	1	7	25·58
2	6	03·62	2	6	65·63	2	7	27·65
3	6	05·68	3	6	67·70	3	7	29·72
4	6	07·75	4	6	69·77	4	7	31·78
5	6	09·82	5	6	71·83	5	7	33·85
6	6	11·89	6	6	73·90	6	7	35·92
7	6	13·95	7	6	75·97	7	7	37·98
8	6	16·02	8	6	78·04	8	7	40·05
9	6	18·09	9	6	80·10	9	7	42·12

Value of Gold per Ounce Troy at different Fineness.

Thousandths Fine	Dollars	Cents	Thousandths Fine	Dollars	Cents	Thousandths Fine	Dollars	Cents
·360	7	44·19	·390	8	06·20	·420	8	68·22
1	7	46·25	1	8	08·27	1	8	70·28
2	7	48·32	2	8	10·34	2	8	72·35
3	7	50·39	3	8	12·40	3	8	74·42
4	7	52·45	4	8	14·47	4	8	76·49
5	7	54·52	5	8	16·54	5	8	78·55
6	7	56·59	6	8	18·60	6	8	80·62
7	7	58·66	7	8	20·67	7	8	82·69
8	7	60·72	8	8	22·74	8	8	84·75
9	7	62·79	9	8	24·81	9	8	86·82
·370	7	64·86	·400	8	26·87	·430	8	88·89
1	7	66·93	1	8	28·94	1	8	90·96
2	7	68·99	2	8	31·01	2	8	93·02
3	7	71·06	3	8	33·07	3	8	95·09
4	7	73·13	4	8	35·14	4	8	97·16
5	7	75·19	5	8	37·21	5	8	99·22
6	7	77·26	6	8	39·28	6	9	01·29
7	7	79·32	7	8	41·34	7	9	03·36
8	7	81·39	8	8	43·41	8	9	05·43
9	7	83·46	9	8	45·48	9	9	07·49
·380	7	85·53	·410	8	47·55	·440	9	09·56
1	7	87·60	1	8	49·61	1	9	11·63
2	7	89·66	2	8	51·68	2	9	13·70
3	7	91·73	3	8	53·75	3	9	15·76
4	7	93·80	4	8	55·81	4	9	17·83
5	7	95·87	5	8	57·88	5	9	19·90
6	7	97·93	6	8	59·95	6	9	21·96
7	8	00·00	7	8	62·02	7	9	24·03
8	8	02·07	8	8	64·08	8	9	26·10
9	8	04·13	9	8	66·15	9	9	28·17

PROCESSES OF SILVER AND GOLD **EXTRACTION**. 317

Value of Gold per Ounce Troy at different Fineness.

Thousandths Fine	Dollars	Cents	Thousandths Fine	Dollars	Cents	Thousandths Fine	Dollars	Cents
·450	9	30·23	·480	9	92·25	·510	10	54·26
1	9	32·30	1	9	94·32	1	10	56·33
2	9	34·37	2	9	96·38	2	10	58·40
3	9	36·43	3	9	98·45	3	10	60·47
4	9	38·50	4	10	00·52	4	10	62·53
5	9	40·57	5	10	02·58	5	10	64·60
6	9	42·64	6	10	04·65	6	10	66·67
7	9	44·70	7	10	06·72	7	10	68·73
8	9	46·77	8	10	08·79	8	10	70·80
9	9	48·84	9	10	10·85	9	10	72·87
·460	9	50·90	·490	10	12·92	·520	10	74·94
1	9	52·97	1	10	14·99	1	10	77·00
2	9	55·04	2	10	17·05	2	10	79·07
3	9	57·11	3	10	19·12	3	10	81·14
4	9	59·17	4	10	21·19	4	10	83·20
5	9	61·24	5	10	23·26	5	10	85·27
6	9	63·31	6	10	25·32	6	10	87·34
7	9	65·37	7	10	27·39	7	10	89·41
8	9	67·44	8	10	29·46	8	10	91·47
9	9	69·51	9	10	31·52	9	10	93·54
·470	9	71·58	·500	10	33·59	·530	10	95·61
1	9	73·64	1	10	35·66	1	10	97·67
2	9	75·71	2	10	37·73	2	10	99·74
3	9	77·78	3	10	39·79	3	11	01·81
4	9	79·84	4	10	41·86	4	11	03·88
5	9	81·91	5	10	43·93	5	11	05·94
6	9	83·98	6	10	45·99	6	11	08·01
7	9	86·05	7	10	48·06	7	11	10·08
8	9	88·11	8	10	50·13	8	11	12·14
9	9	90·18	9	10	52·20	9	11	14·21

Value of Gold per Ounce Troy at different Fineness.

Thousandths Fine	Dollars	Cents	Thousandths Fine	Dollars	Cents	Thousandths Fine	Dollars	Cents
·540	11	16·28	·570	11	78·29	·600	12	40·31
1	11	18·35	1	11	80·36	1	12	42·38
2	11	20·41	2	11	82·43	2	12	44·44
3	11	22·48	3	11	84·50	3	12	46·51
4	11	24·55	4	11	86·56	4	12	48·58
5	11	26·61	5	11	88·63	5	12	50·65
6	11	28·68	6	11	90·70	6	12	52·71
7	11	30·75	7	11	92·76	7	12	54·78
8	11	32·82	8	11	94·83	8	12	56·85
9	11	34·88	9	11	96·90	9	12	58·91
·550	11	36·95	·580	11	98·97	·610	12	60·98
1	11	39·02	1	12	01·03	1	12	63·05
2	11	41·09	2	12	03·10	2	12	65·12
3	11	43·15	3	12	05·17	3	12	67·18
4	11	45·22	4	12	07·24	4	12	69·25
5	11	47·29	5	12	09·30	5	12	71·32
6	11	49·35	6	12	11·37	6	12	73·39
7	11	51·42	7	12	13·44	7	12	75·45
8	11	53·49	8	12	15·50	8	12	77·52
9	11	55·56	9	12	17·57	9	12	79·59
·560	11	57·62	·590	12	19·64	·620	12	81·65
1	11	59·69	1	12	21·71	1	12	83·72
2	11	61·76	2	12	23·77	2	12	85·79
3	11	63·82	3	12	25·84	3	12	87·86
4	11	65·89	4	12	27·91	4	12	89·92
5	11	67·96	5	12	29·97	5	12	91·99
6	11	70·03	6	12	32·04	6	12	94·06
7	11	72·09	7	12	34·11	7	12	96·12
8	11	74·16	8	12	36·18	8	12	98·19
9	11	76·23	9	12	38·24	9	13	00·26

PROCESSES OF SILVER AND GOLD EXTRACTION. 319

Value of Gold per Ounce Troy at different Fineness.

Thousandths Fine	Dollars	Cents	Thousandths Fine	Dollars	Cents	Thousandths Fine	Dollars	Cents
·630	13	02·33	·660	13	64·34	·690	14	26·36
1	13	04·39	1	13	66·41	1	14	28·42
2	13	06·46	2	13	68·48	2	14	30·49
3	13	08·53	3	13	70·54	3	14	32·56
4	13	10·59	4	13	72·61	4	14	34·63
5	13	12·66	5	13	74·68	5	14	36·69
6	13	14·73	6	13	76·74	6	14	38·76
7	13	16·80	7	13	78·81	7	14	40·83
8	13	18·86	8	13	80·88	8	14	42·89
9	13	20·93	9	13	82·95	9	14	44·96
·640	13	23·00	·670	13	85·01	·700	14	47·03
1	13	25·06	1	13	87·08	1	14	49·10
2	13	27·13	2	13	89·15	2	14	51·16
3	13	29·20	3	13	91·21	3	14	53·23
4	13	31·27	4	13	93·28	4	14	55·30
5	13	33·33	5	13	95·35	5	14	57·36
6	13	35·40	6	13	97·42	6	14	59·43
7	13	37·47	7	13	99·48	7	14	61·50
8	13	39·53	8	14	01·55	8	14	63·57
9	13	41·60	9	14	03·62	9	14	65·63
·650	13	43·67	·680	14	05·68	·710	14	67·70
1	13	45·74	1	14	07·75	1	14	69·76
2	13	47·80	2	14	09·82	2	14	71·83
3	13	49·87	3	14	11·89	3	14	73·90
4	13	51·93	4	14	13·95	4	14	75·97
5	13	54·01	5	14	16·02	5	14	78·04
6	13	56·07	6	14	18·09	6	14	80·10
7	13	58·14	7	14	20·16	7	14	82·17
8	13	60·21	8	14	22·22	8	14	84·24
9	13	62·27	9	14	24·29	9	14	86·30

320   PROCESSES OF SILVER AND GOLD EXTRACTION.

Value of Gold per Ounce Troy at different Fineness.

Thousandths Fine	Dollars	Cents	Thousandths Fine	Dollars	Cents	Thousandths Fine	Dollars	Cents
·720	14	88·37	·750	15	50·39	·780	16	12·40
1	14	90·44	1	15	52·45	1	16	14·47
2	14	92·51	2	15	54·52	2	16	16·54
3	14	94·57	3	15	56·59	3	16	18·60
4	14	96·64	4	15	58·66	4	16	20·67
5	14	98·71	5	15	60·72	5	16	22·74
6	15	00·78	6	15	62·79	6	16	24·81
7	15	02·84	7	15	64·86	7	16	26·87
8	15	04·91	8	15	66·93	8	16	28·94
9	15	06·98	9	15	68·99	9	16	31·01
·730	15	09·04	·760	15	71·06	·790	16	33·07
1	15	11·11	1	15	73·13	1	16	35·14
2	15	13·18	2	15	75·19	2	16	37·21
3	15	15·25	3	15	77·26	3	16	39·28
4	15	17·31	4	15	79·33	4	16	41·34
5	15	19·38	5	15	81·40	5	16	43·41
6	15	21·45	6	15	83·46	6	16	45·48
7	15	23·51	7	15	85·53	7	16	47·55
8	15	25·58	8	15	87·60	8	16	49·61
9	15	27·65	9	15	89·66	9	16	51·68
·740	15	29·72	·770	15	91·73	·800	16	53·75
1	15	31·78	1	15	93·80	1	16	55·81
2	15	33·85	2	15	95·87	2	16	57·88
3	15	35·92	3	15	97·93	3	16	59·95
4	15	37·98	4	16	00·00	4	16	62·02
5	15	40·05	5	16	02·07	5	16	64·08
6	15	42·12	6	16	04·13	6	16	66·15
7	15	44·18	7	16	06·20	7	16	68·22
8	15	46·25	8	16	08·27	8	16	70·28
9	15	48·32	9	16	10·34	9	16	72·35

PROCESSES OF SILVER AND GOLD EXTRACTION. 321

Value of Gold per Ounce Troy at different Fineness.

Thousandths Fine.	Dollars.	Cents.	Thousandths Fine.	Dollars.	Cents.	Thousandths Fine.	Dollars.	Cents.
·810	16	74·42	·840	17	36·43	·870	17	98·45
1	16	76·49	1	17	38·50	1	18	00·52
2	16	78·55	2	17	40·57	2	18	02·58
3	16	80·62	3	17	42·64	3	18	04·65
4	16	82·69	4	17	44·70	4	18	06·72
5	16	84·75	5	17	46·77	5	18	08·79
6	16	86·82	6	17	48·84	6	18	10·85
7	16	88·89	7	17	50·90	7	18	12·92
8	16	90·96	8	17	52·97	8	18	14·99
9	16	93·02	9	17	55·04	9	18	17·05
·820	16	95·09	·850	17	57·11	·880	18	19·12
1	16	97·16	1	17	59·17	1	18	21·19
2	16	99·22	2	17	61·24	2	18	23·26
3	17	01·29	3	17	63·31	3	18	25·32
4	17	03·36	4	17	65·37	4	18	27·39
5	17	05·43	5	17	67·44	5	18	29·46
6	17	07·49	6	17	69·51	6	18	31·52
7	17	09·56	7	17	71·58	7	18	33·59
8	17	11·63	8	17	73·64	8	18	35·66
9	17	13·70	9	17	75·71	9	18	37·73
·830	17	15·76	·860	17	77·78	·890	18	39·79
1	17	17·83	1	17	79·84	1	18	41·86
2	17	19·90	2	17	81·91	2	18	43·93
3	17	21·96	3	17	83·98	3	18	45·99
4	17	24·03	4	17	86·05	4	18	48·06
5	17	26·10	5	17	88·11	5	18	50·13
6	17	28·17	6	17	90·18	6	18	52·20
7	17	30·23	7	17	92·25	7	18	54·26
8	17	32·30	8	17	94·32	8	18	56·33
9	17	34·37	9	17	96·38	9	18	58·40

Value of Gold per Ounce Troy at different Fineness.

Thousandths Fine	Dollars	Cents	Thousandths Fine	Dollars	Cents	Thousandths Fine	Dollars	Cents
·900	18	60·46	·930	19	22·48	·960	19	84·50
1	18	62·53	1	19	24·55	1	19	86·56
2	18	64·60	2	19	26·61	2	19	88·63
3	18	66·67	3	19	28·68	3	19	90·70
4	18	68·73	4	19	30·75	4	19	92·76
5	18	70·80	5	19	32·82	5	19	94·83
6	18	72·87	6	19	34·88	6	19	96·90
7	18	74·94	7	19	36·95	7	19	98·97
8	18	77·00	8	19	39·02	8	20	01·03
9	18	79·07	9	19	41·08	9	20	03·10
·910	18	81·14	·940	19	43·15	·970	20	05·17
1	18	83·20	1	19	45·22	1	20	07·23
2	18	85·27	2	19	47·29	2	20	09·30
3	18	87·34	3	19	49·35	3	20	11·37
4	18	89·41	4	19	51·42	4	20	13·44
5	18	91·47	5	19	53·49	5	20	15·50
6	18	93·54	6	19	55·56	6	20	17·57
7	18	95·61	7	19	57·62	7	20	19·64
8	18	97·67	8	19	59·69	8	20	21·70
9	18	99·74	9	19	61·76	9	20	23·77
·920	19	01·81	·950	19	63·82	·980	20	25·84
1	19	03·88	1	19	65·89	1	20	27·91
2	19	05·94	2	19	67·96	2	20	29·97
3	19	08·01	3	19	70·03	3	20	32·04
4	19	10·08	4	19	72·09	4	20	34·11
5	19	12·14	5	19	74·16	5	20	36·18
6	19	14·21	6	19	76·23	6	20	38·24
7	19	16·28	7	19	78·29	7	20	40·31
8	19	18·35	8	19	80·36	8	20	42·38
9	19	20·41	9	19	82·44	9	20	44·44

PROCESSES OF SILVER AND GOLD EXTRACTION. 323

Value of Gold per Ounce Troy at different Fineness.

Thousandths Fine	Dollars	Cents	Thousandths Fine	Dollars	Cents	Thousandths Fine	Dollars	Cents
·990	20	46·51	·994	20	54·78	·998	20	63·05
1	20	48·58	5	20	56·85	9	20	65·12
2	20	50·65	6	20	58·91	1000	20	67·18
3	20	52·71	7	20	60·98			

INDEX.

	Page.
Abstrich	238
Abzug	238
Acetate of lead	192
Acid, sulphuric	71
Acid, hydrochloric	96
Agitator	120, 173
Alum	71
Amalgamation of gold	59
Amalgamation of gold in arrastras	61
Amalgamation of gold **in batteries**	59
Amalgamation of gold in **pans**	63
Amalgamation of silver	261
Amalgamation of silver in barrels	117, 262
Amalgamation of silver in pans	76, 124
Amalgamation **of silver** in Veatch's Tubs	122
Amalgamation of silver in Wheeler's Pans	81
Amalgamation of copper matt	267
Amalgamation of speiss	268
Amalgamation of black copper	268
Analysis of amalgam and bullion metal	26
Annealing of crucibles	135
Antimonial blend	41
Antimonial silver	45
Argentiferous copper ore	190
Argentiferous gray copper ore	42
Argentiferous lead ores	190
Argentiferous pyrites	191
Argentiferous zinc ores	191
Arsenical blend	42
Arquirite	47
Assay of silver with the blowpipe	28, 199
Assay of silver **by fire**	48, 192
Assay of silver and gold	50
Assay of lead	57
Assay of rich silver ores	193
Assay of silver with lead	193

	Page.
Assay of silver without lead	195
Assay of silver with litharge or acetate of lead	195
Assay of roasted **ores**	196
Assay of poor ores	196
Assay of ores rich in **sulphurets**	196
Assay of ores rich in earths	196
Assay of alloys	197
Assay of silver and lead	197
Assay of silver, tin, and zinc	197
Assay of silver, copper, and **brass**	198
Assay of cupriferous silver	198
Assay in the wet way	199
Assay for matt	227
Augustin's Process	270
Bismuth silver	47
Bisulphate of soda	11
Blast furnaces	146
Blend, antimonial	41
Blend, arsenical	42
Blowpipe, use of	15
Borax	10
Borax glass	11
Brightening	241
Brittle silver ore	40
Bromic **silver**	44
Bromyrite	44
Calomel	106
Chemical **action**	265
Chemicals	69, 83
Chemicals per ton of ore	74
Chloride of silver	43
Chlorination, of gold ores	64
Chlorine, action on metals	65, 94
Chlorobromide of **silver**	44
Coating, on charcoal	17
Concentration of silver in lead	164, 244
Concentration of silver in zinc	249

326 INDEX.

	Page.
Consume, of quicksilver	130
Copperas	71
Copper, black	231
Copper, chloride of	72
Copper dissolving process	234
Copper, oxyd of	12
Copper, subchloride of	73
Copper, sulphate of	69
Crucible furnace	179
Crucible cast iron	259
Cupels	49
Cupels, size of	198
Cupel mass	184, 236
Cupellation with blowpipe	31
Cupellation under muffle	53, 194
Cupellation on hearth	154, 235
Cupelling furnace	183
Dark red silver ore	41
Division of silver ores	189
Dressing of assay samples	199
Embolite	44
Eucairite	46
Eugen-glance	40
Examination of ores for roasting	98
Examination on charcoal	16
Examination with soda and borax	18
Examination in a closed glass tube	20
Examination in an open glass tube	22
Extraction of gold	59
Extraction of silver, methods	205
Extraction of silver in the dry way	216
Extraction of silver with lead	216
Extraction of silver from lead with zinc	249
Extraction of silver in the wet way	261
Extraction of silver by precipitation	270
Extraction of silver from copper matt.	270
Filtering apparatus	274
Flame	15
Fluxes for melting ores	148
Froth	155, 238, 242
Furnace for assays	52
Furnace for cupellation	183
Furnace for meting bars	135, 179
Furnace for melting ores	180
Furnace for refining silver	186
Furnace for roasting ores	176, 177
Galvanic action	266
Gold	36
Gold with mercury	37
Gold with silver	36
Gold with tellurium	37
Gold with tellurium and lead	37

	Page.
Granulation of lead	149
Hardness of silver ores	35
Hearth	243
Hessite	46
Horn silver	43
Hydrostatic melting	229
Hyposulphite of soda	279
Incorporation	131
Iodyrite	44
Iodide of silver and mercury	45
Iron, protochlorid of	73
Iron, chlorid of	74
Iron, as chemical	84
Lead, for cupriferous assays	198, 200
Light red silver ore	42
Limadura	131
Liquation	231
Litharge	148
Litharge ring	157
Litharge fumes	158
Litharge black	238
Litharge red	242
Litharge yellow	242
Lixiviation	273
Loss of lead, remedy for	218
Loss of silver, remedy for	258
Loss of silver in roasting	115
Lustre of ores	24
Magistral	129
Marl	184, 236
Matt	136, 226
Melting of retorted amalgam.	134, 138, 139
Melting of silver ores with lead	205
Melting of rich ores in crucibles	220
Melting in cupelling furnaces	221
Melting of unroasted ores	222
Melting of roasted argentiferous copper ores	224
Melting of rich silver ores	152
Melting furnace	180
Melting process	145
Methods of extraction, choice of	218
Methods, principal	211
Methods of refining	254
Miargyrite	41
Mixture of ore for melting	151, 222, 223
Naumannite	46
Oxydizing flame	15
Pans, description of	169, 170, 174
Parke's process for desilverising lead	249
Patera's process for extraction of silver	279

	Page.
Patio, or American heap amalgamation	164, 244
Pattinson's concentration of silver in lead	164, 244
Polybasite	40
Polysulphide of sodium	279
Proustite	42
Protochloride of iron	73
Purification of lead	167
Pyrargyrite	41
Reagents for blowpipe	10, 12
Reduction flame	15
Reduction of litharge	161
Refining of silver	162, 251
Refining in crucibles	162, 259
Refining in hearth furnaces	163
Refining on movable tests	254
Refining under muffles	255
Refining in reverberatory furnaces	257
Refining furnace	186
Retorting	133
Roasting	90
Roasting for barrel amalgamation	101, 263
Roasting for pan amalgamation	106
Roasting for patio process	129
Roasting for Augustin's process	271
Roasting for Ziervogel's process	276
Roasting for Patera's process	280
Roharbeit	225
Rohstein	225
Ruby silver (see pyrargyrite)	41
Salt, common	72, 96, 130
Scheme of Pattinson's crystallization	248
Scorification, assay	193
Scrapings	155, 238
Selenid of silver	46
Selenid of silver and copper	46
Separation of lead and silver amalgam	113
Sign of completed roasting	277
Silver amalgam	47
Silver assay	192
Silver bromic	44
Silver, chlorid of	43
Silver chlorobromid	44
Silver copper glance	39

	Page.
Silver fahlerz	42
Silver glance	38
Silver, iodid of	44
Silver native	38
Silver ores	38
Silver, sulphuret of	38
Silver, tellurid of	46
Silver concentration in matt	225
Silver concentration in lead	164, 244
Silver concentration in zinc	249
Silver extraction by mercury	207
Silver extraction by precipitation	208
Silver extraction by lead	228
Skimmer	136
Soda bisulphate	11, 71
Soda carbonate	11, 49
Specific gravity	33
Spitting of silver	253
Spitting of silver, remedy for	257
Steam application in roasting	100
Sternbergite	39
Stromeyerite	39
Sublimation in glass tubes	21
Sulphuret of silver and iron	39
Systematic proceeding in determining gold and silver ores	23
Table, loss of silver by cupellation	33
Table, progress of enriching lead	165
Tailings	80, 89, 108
Temperature in cupelling	157, 258
Tellurid of silver	46
Test-ring	243
Tin foil	12
Torta	131
Treatment of slag and matt from melting bars	141
Treatment of rich silver ores	125, 209
Treatment of poor silver ores	210
Value of gold and silver per ounce	145
Vitriol blue	69
Vitriol green	71
Volatility of silver	100
Weights, for assay	13
Wet assay	199
Wet process	67
Xanthocone	42

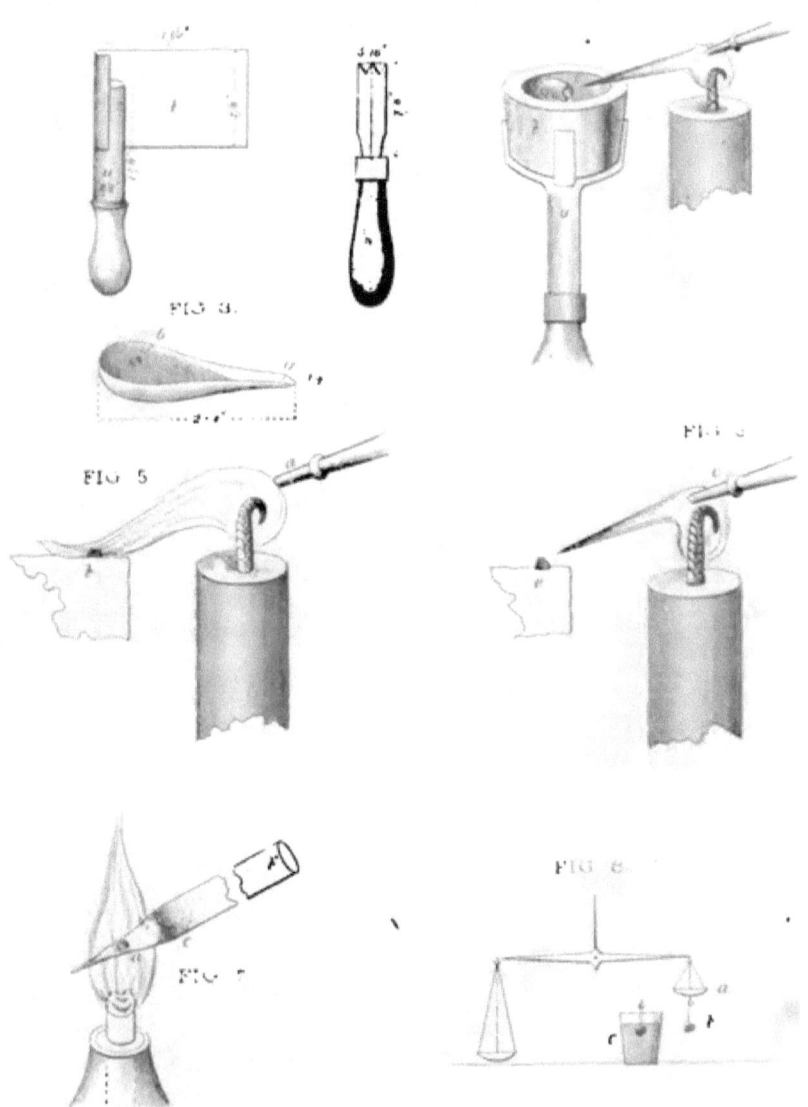

FIG. 9.

FIG. 10.

FIG. 11.

FIG. 12.

FIG. 13.

FIG. 15.

FIG. 16.

FIG. 14.

Fishbourne's Lithog. S F

Scale for Figs 13. 14. 15 inch to the foot

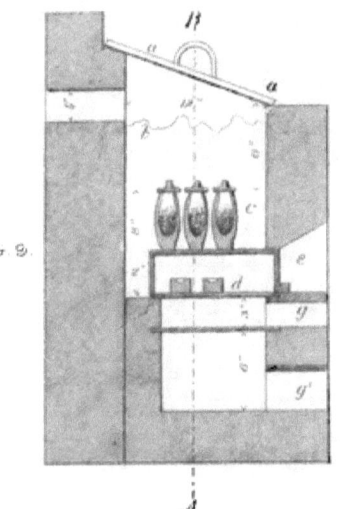

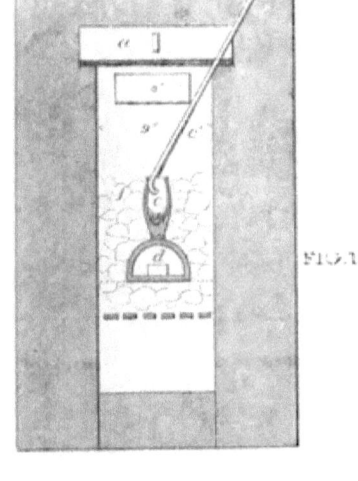

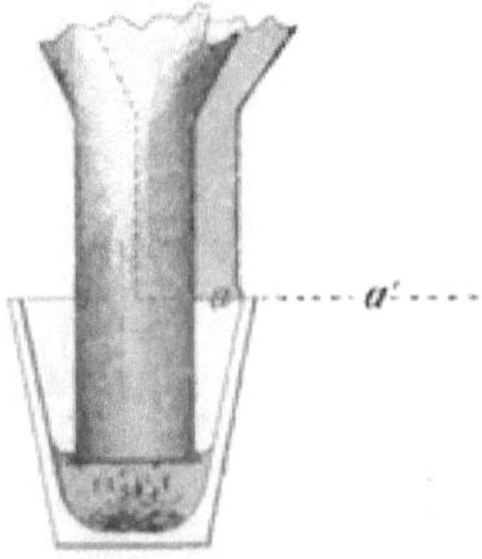

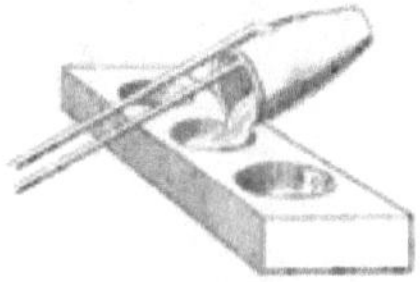

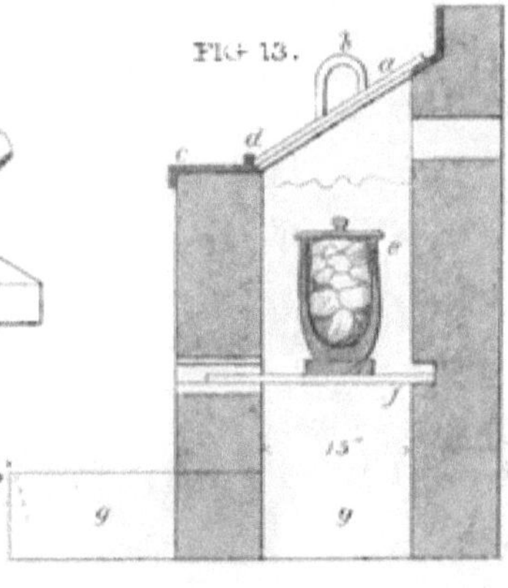

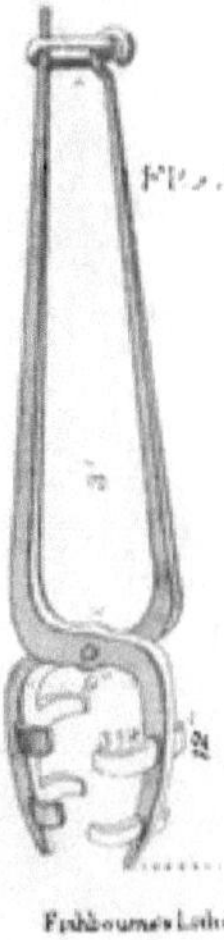

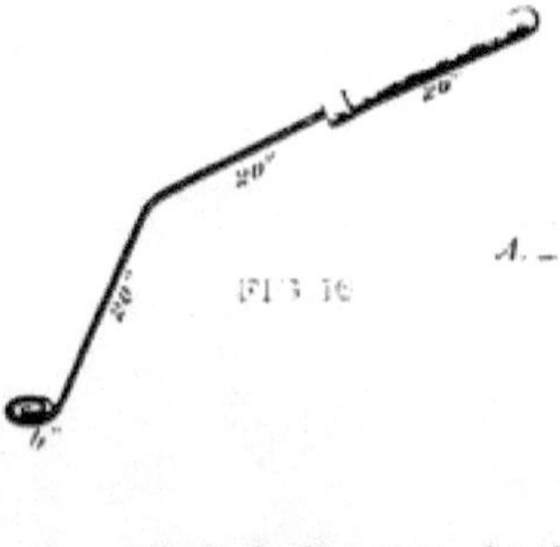

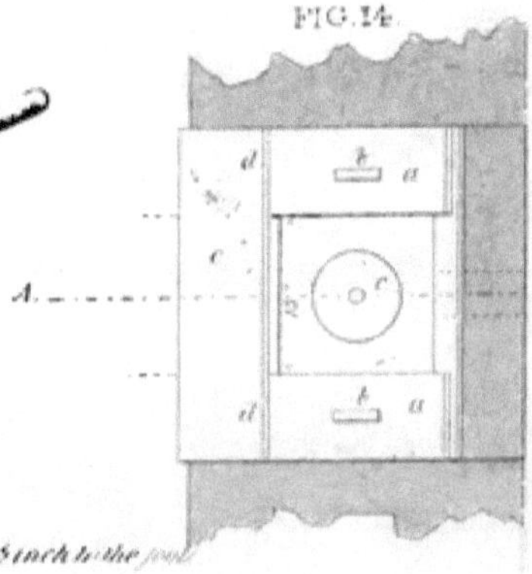

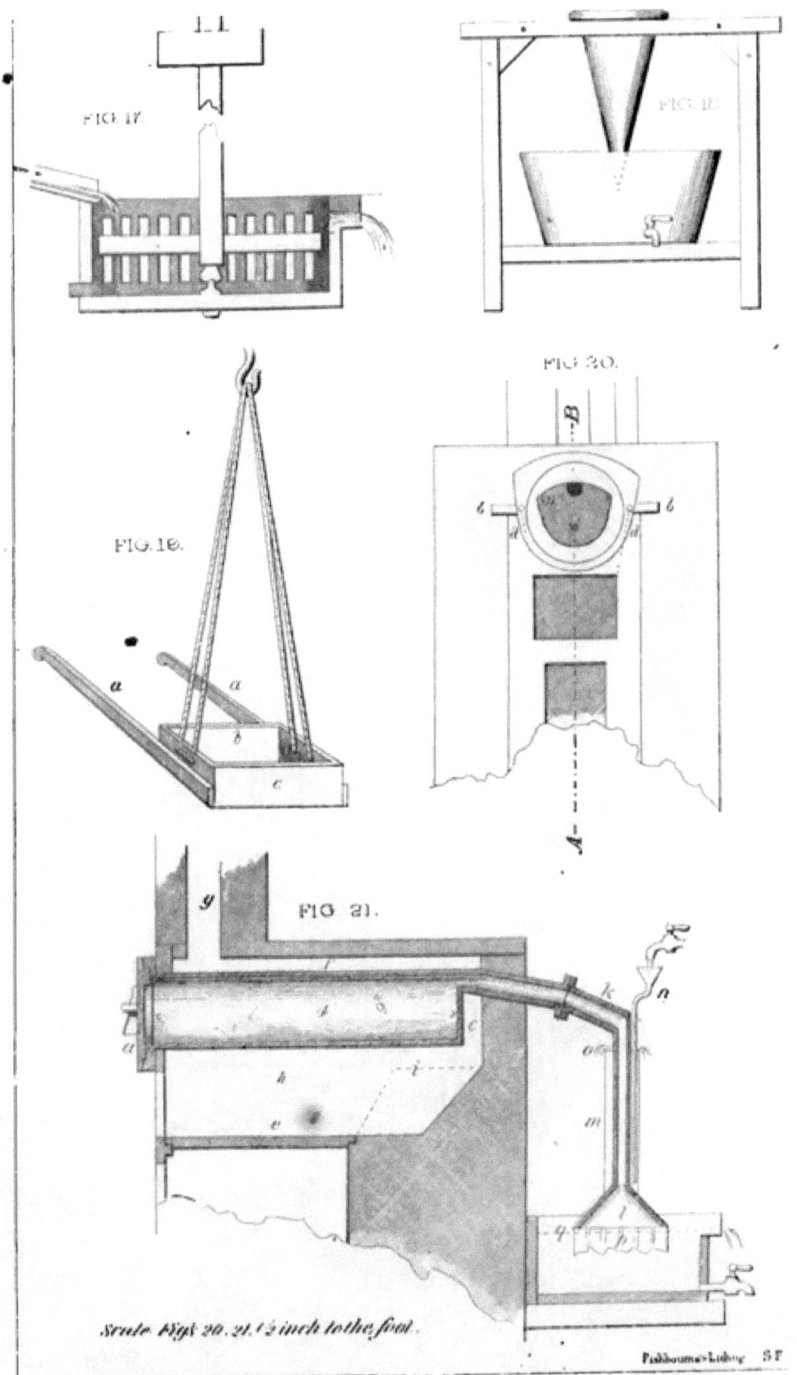

FIG 22.

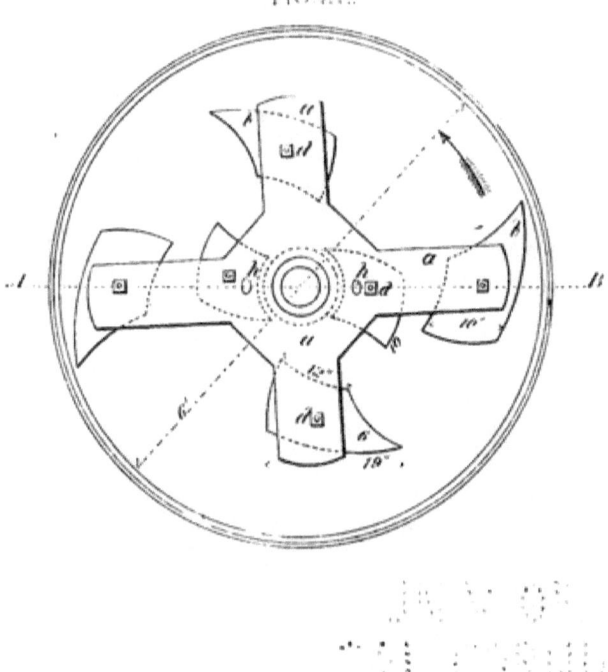

FIG 23

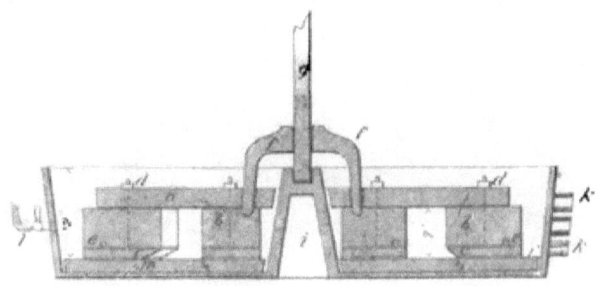

Scale Figs 22, 23 ½ inch to the foot.

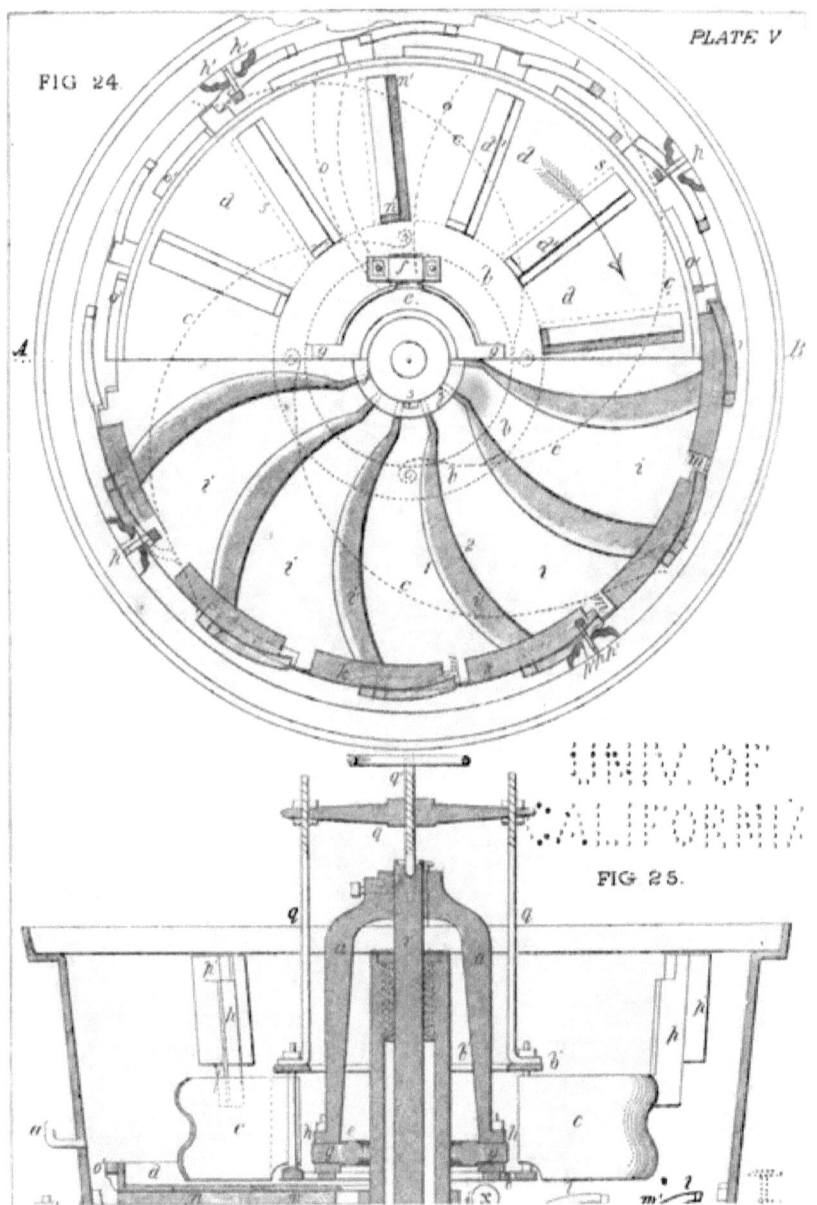

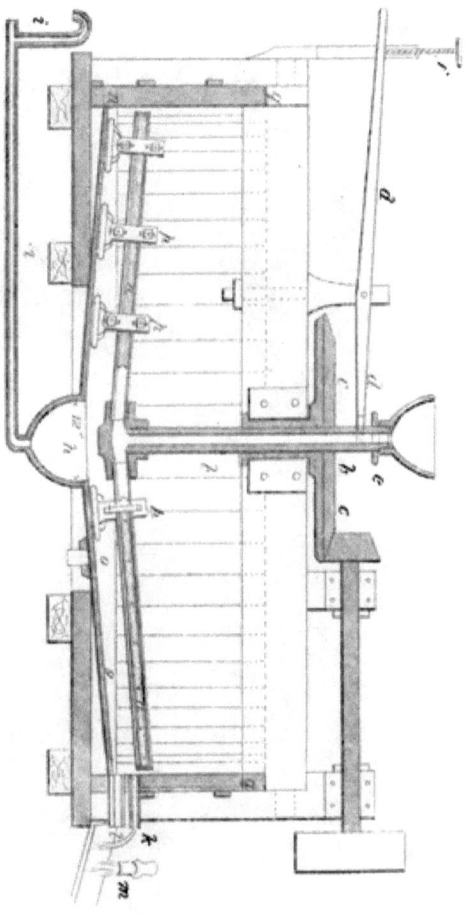

Scale ½ inch to the foot.

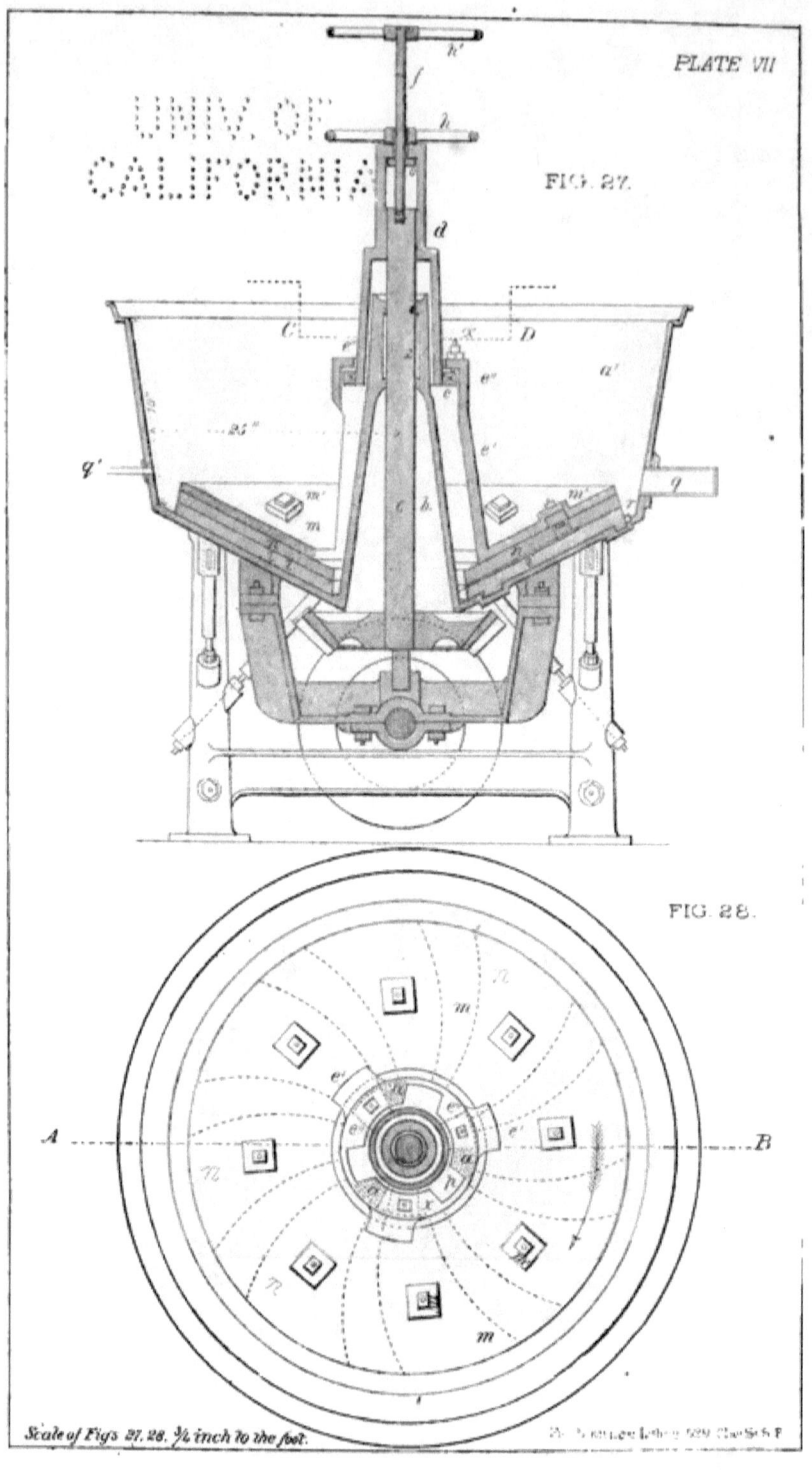

PLATE VII.

FIG. 27.

FIG. 28.

Scale of Figs. 27, 28. ½ inch to the foot.

FIG. 29.

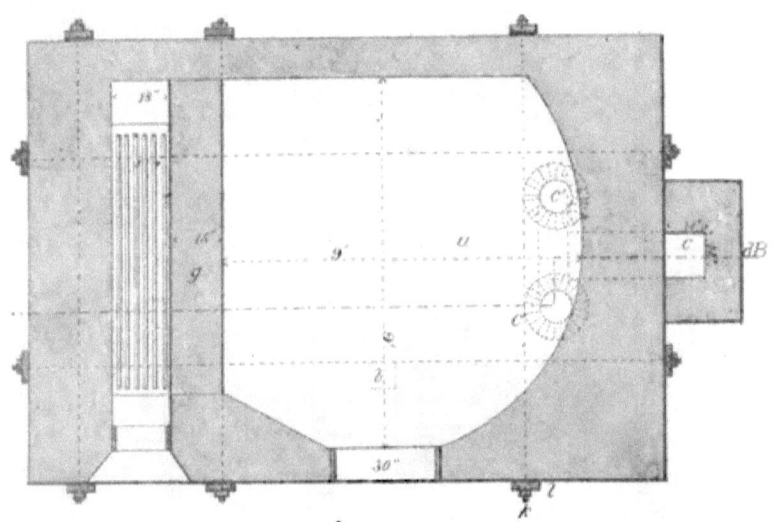

FIG. 30.

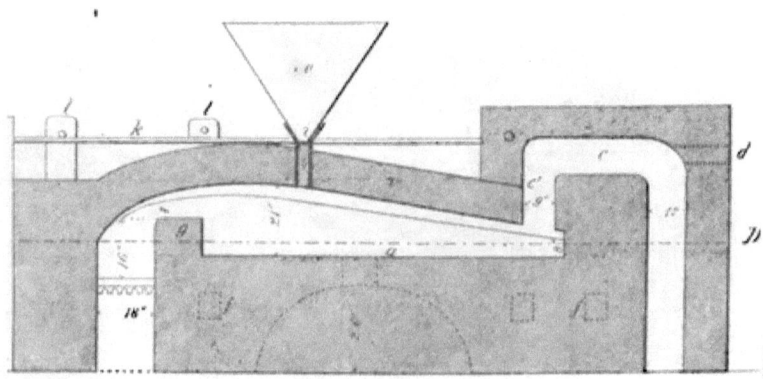

Scale Fig's 29..30 ¼ inch to the foot.

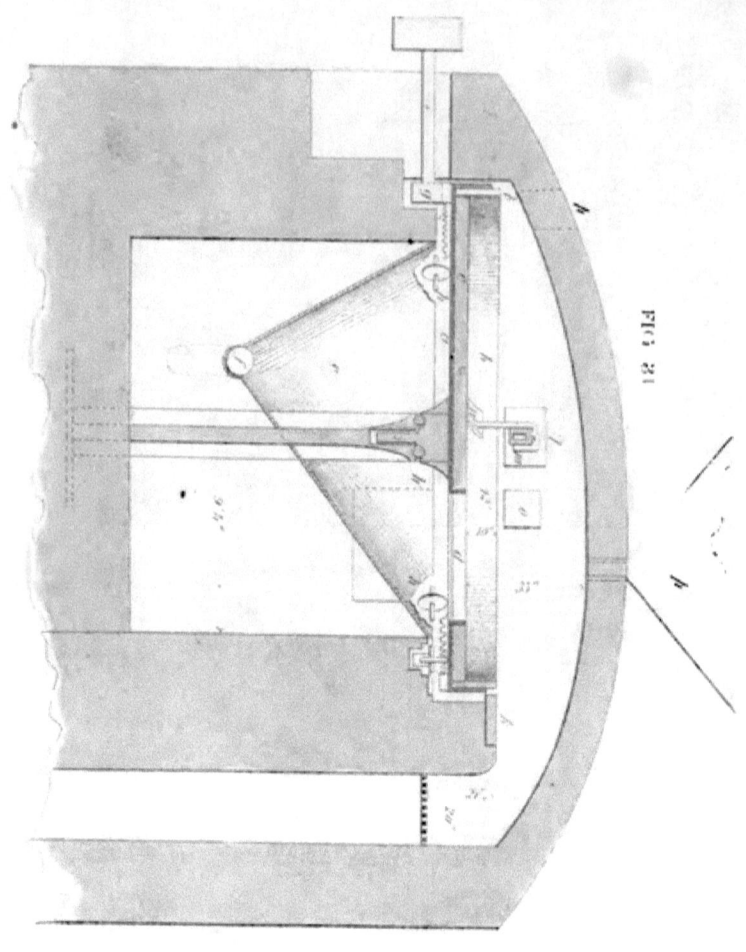

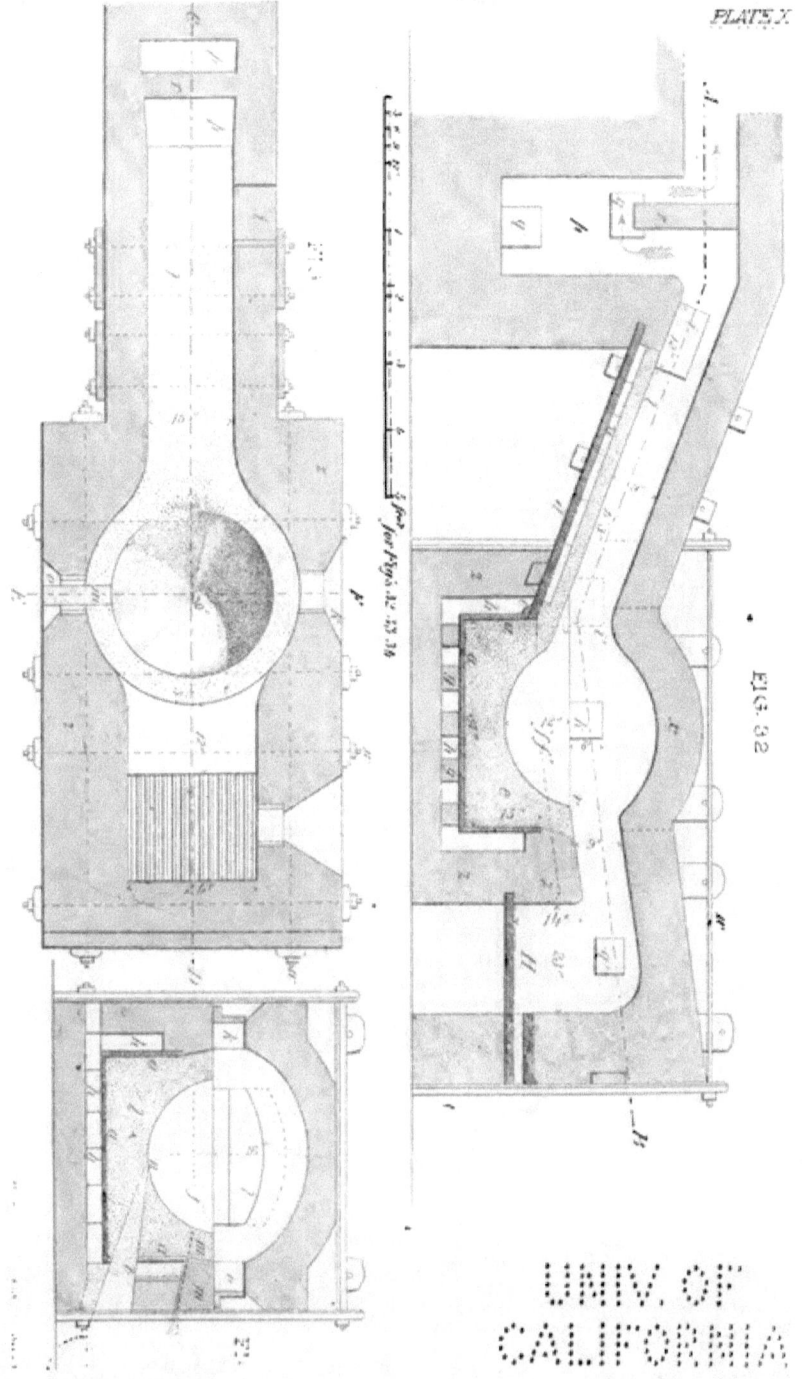

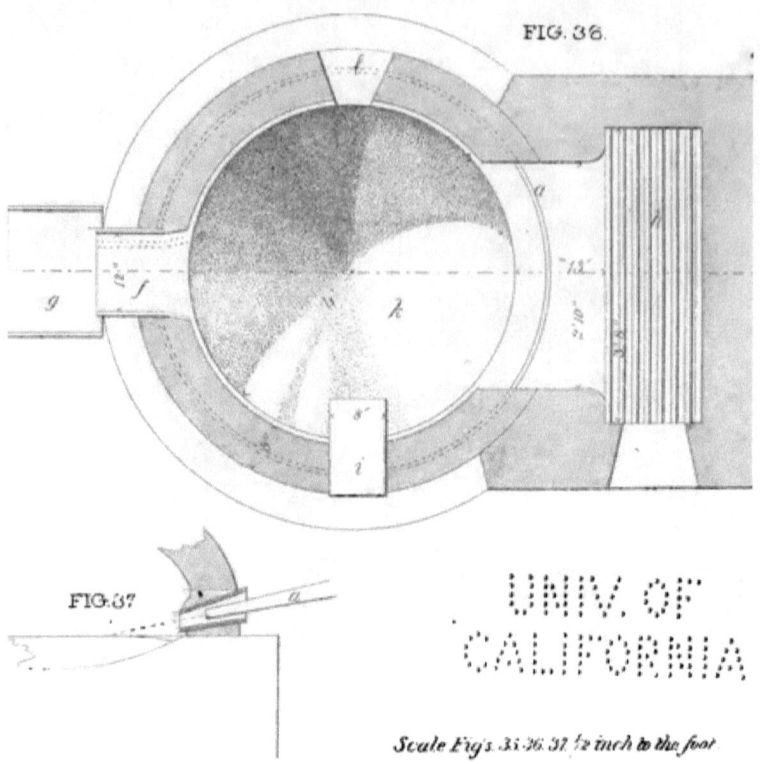

www.ingramcontent.com/pod-product-compliance
Lightning Source LLC
Chambersburg PA
CBHW032048220426
43664CB00008B/909